11 Crystals

Growth, Properties, and Applications

Managing Editor: H. C. Freyhardt

Superhard Materials, Convection, and Optical Devices

With Contributions by

R. B. Heimann   J. Kleiman   D. Schwabe
H.-J. Weber

Springer-Verlag
Berlin Heidelberg New York
London Paris Tokyo

ISBN-13:978-3-642-73207-2 e-ISBN-13:978-3-642-73205-8
DOI: 10.1007/978-3-642-73205-8

Softcover reprint of the hardcover 1st edition 1988

Typesetting: Mitterweger Werksatz GmbH, 6831 Plankstadt, Germany

2152/3140-543210

Table of Contents

Shock-Induced Growth of Superhard Materials

R. B. Heimann* and J. Kleiman

3M Canada Inc. Corporate R & D, 4925 Dufferin Street, Downsview, Ontario, M3H 5T6, Canada

The use of superhard materials, particularly diamond, for a broad variety of industrial operations such as cutting, drilling, grinding, lapping, and polishing of hardened steel, hard alloys or rock has become so widespread that the demand cannot be met anymore from natural sources. Consequently, considerable effort has been expended to synthesize those materials by static or dynamic high-pressure techniques during the last thirty years. Elevated temperatures, however, promote oxidation of diamond or formation of carbides with the metal to be processed. Thus, increasing use is made of the diamond-like form of boron nitride for purposes where high machining speed is required.

Dynamic high-pressure techniques utilizing shock waves generated by very sudden release of mechanical, chemical, electrical, or radiation energy in a confined space are frequently used to transform graphite or the graphite-like form of boron nitride into diamond or cubic boron nitride ("borazon"), respectively.

Although the very short duration of a shock wave does not allow for attainment of equilibrium conditions in the material to be transformed, high-density phases can be synthesized readily in an orderly fashion by phase transformation within 10^{-7} to 10^{-9} s. Strong unidirectional shear stresses generated by the shock wave may provide an additional driving force for the phase transformation.

Once nuclei of the new phase have been formed by suitable timing of high-temperature and high-pressure pulses, care has to be taken to avoid reconversion of the high-pressure phase in the wake of the rarefaction wave at prevailing high post-shock temperatures. Thus, shock-quenching by adiabatic expansion of the compressed matter appears to be the key feature to ensure high yield of diamond or cubic boron nitride.

* Present address: Alberta Research Council, Edmonton, Alberta T6H 5X2

"... diamond can form in several different ways, and stubborn mysteries still surround some of them."

Bovenkerk, H. P., et al.
Nature *184,* 1094 (1959)

"Although a great deal has been learned about (elemental carbon) during the present century, it is obvious that there is much more to be learned."

Bundy, F. P., J. Geophys. Research
85 (B12), 6930 (1980)

1 Introduction

Superhard abrasive materials such as diamond and boron nitride have important industrial applications as tools for cutting, drilling, grinding and polishing utilized to machine and finish hardened steel and other hard alloys, as well as drilling metal and rock. The use of superhard materials, particularly diamond, for a broad variety of industrial operations has become so widespread that the demand cannot be met from natural sources. Consequently, considerable efforts have been expended to synthesize diamond by dynamic high-pressure techniques (see for example[1–5]). Diamond, however, tends to react with oxygen at much lower temperatures than boron nitride, and also reacts with transition metals at elevated temperatures to form carbides. Therefore, much work has been devoted to synthesize high-pressure forms of boron nitride (see for example[6–9]).

The first successful diamond synthesis employing shock waves was achieved in 1961 by DeCarli and Jamieson[1] and, almost contemporaneously by Alder and Christian[10]. These authors subjected compressed graphite powder to shock pressures of about 30 GPa (300 kbar), and a few percent of the graphite were converted to very-fine grained cubic diamond. Synthesis of boron nitride with hexagonal (wurtzite) structure (w-BN) was first reported by Adadurov et al.[6] in 1967 as the result of shock compression of graphite-like boron nitride (g-BN) at dynamic pressures exceeding 12.8 GPa. The cubic (zincblende) modification of boron nitride (z-BN) was first detected in shock recovery experiment by Batsanov et al.[11] in 1965. Since then, much work has been carried out directed predominantly towards increase of crystal size, homogeneity and yield of the high-pressure phases as well as optimization of the experimental designs used to generate shock pressures, and determination of the linear shock-wave velocity – particle velocity relationship (Rankine-Hugoniot)[12] of the materials involved.

Shock wave experiments were originally performed to extend the range of pressure-volume data beyond the region that could be reached with conventional static pressure experiments. Shock data, however, do not extend to low pressures[1].

1 If the linear Hugoniot relation $U_s = a + b \cdot U_p$
(U_s = shock velocity, U_p = particle velocity, a and b constants) holds down to zero pressure (= zero particle velocity, U_p), the constant "a" becomes identical to the hydrodynamic sound velocity. Thus, shock and sound speed become identical at zero particle velocity[13]

Table 1 lists some properties that static and dynamic compression impose on matter and on the physical conditions of the experiment. The most striking difference between the two kinds of experiments is the length of time a sample is being subjected to high pressure. Moreover, shear stress in statically compressed samples is usually weak and randomly distributed, but is strong in shock-compressed samples and also unidirectional. Those shear stresses are thought to provide an additional driving force for phase transformations, and in that way make up for the short interaction time.

The short interaction time of a shock wave with matter (nano- to microseconds) generally does not allow for attainment of equilibrium conditions that are frequently present in the static pressure regime. However, comparison of results obtained by hydrostatic pressure loading and shock-loading often yields remarkable agreement if corrections are made for shear introduced by the one-dimensional strain in shock wave experiments[14–16].

The interaction of shock waves and matter results generally in an irreversible lattice compression that in many cases is accompanied by a phase transformation. It is this phase transformation that led to the application of shock loading techniques to the carbon and boron nitride systems to generate the industrially applicable superhard materials (cubic) diamond, and hexagonal (w-BN) or cubic boron nitride (z-BN).

It was thought that the transformation of the graphite-like forms of carbon and boron nitride, respectively, into the zincblende-like forms (diamond, z-BN) proceeds via a diffusionless mechanism involving simple compression of an intermediately formed rhombohedral polytype[1, 17], perhaps, mediated by a dislocation avalanche sweeping through the crystal volume to be converted[18, 19]. Although this straightforward approach had to be modified in the light of more recent experimental findings (see 4.3), it provides a rationale for the apparently instantaneous phase transformation occurring during the extremely short duration of the pressure pulse in shock wave experiments.

There is indication, however, that diffusion-controlled phase transformations do indeed occur in shock wave experiments, involving diffusion coefficients of carbon in shocked iron as high as 10^4 cm^2/s[20]. Unusually high crystal growth rates should be expected from the high diffusion rates and, indeed, growth rates of crystals of diamond and copper bromide as high as 0.3–1 m/s were observed[1, 21].

Table 1. Properties of static and dynamic compression

Property	Static	Dynamic
Duration	Long (seconds to hours)	Ultrashort (nano – to microseconds)
Pressure	Maximum pressure ($\leqq$ 50 GPa) limited by tensile strength of container material.	Maximum pressure (< 2 TPa at present) theoretically unlimited.
Mode of compression	Hydrostatic (uniform stress distribution).	Unidirectional strain.
Shear stress	Weak, randomly distributed.	Strong, one-dimensional.
Temperature	Maximum temperature (~ 2000 °C) limited by thermal properties of container material.	Maximum temperature generally very high, produced by thermodynamic processes.
Parameter control	Good T control, satisfactory P control.	Poor T control, poor P control.

2 Physics of Shock Waves

A shock wave by definition is a high-amplitude stress wave. Mathematically, it is described by a set of hyperbolic differential equations, similar to those describing normal elastic and acoustic waves. For a shock wave, however, the equations are non-linear and contain entropic changes which in turn give rise to irreversible physical effects in solids such as phase transitions, energy band compression, and change of magnetic properties.

The velocity of propagation of a shock wave, U_s, is supersonic with respect to the initial state, $U_s^2 > C_0^2$, where C_0 is the wave velocity of sound in vacuum. The matter in shocked state is usually compressed to a higher density, and simultaneously accelerated to a particle velocity U_p. The displacement velocity behind the advancing shock front is greater than the propagation velocity of the shock itself ($U_p + C_0 > U_0$). Behind the shock front, there is a region of decreasing pressure called "rarefaction wave" which gradually overtakes the shock and in turn leads to its attenuation.

Propagation velocity of the shock wave, U_s and the particle velocity, U_p, are linked by a set of conservation equations:

Conservation of mass:

$$\varrho_0(U_s - U_{p0}) = \varrho(U_s - U_p) , \qquad (1)$$

Conservation of momentum:

$$P - P_0 = \varrho_0(U_s \cdot U_p) , \qquad (2)$$

Conservation of energy (Bernoulli's theorem):

$$P \cdot U_p = \varrho_0 \cdot U_s(E - E_0 + U_p^2/2) , \qquad (3)$$

where U_{p0} = particle velocity ahead of the shock front, P and P_0 = pressure (amplitude) behind and ahead of the shock front, and ϱ and ϱ_0 = material density behind and ahead of the shock front. The terms E and E_0 refer to the specific internal energy of the shocked material behind and ahead of the shock front, respectively.

The conservation Eqs. (1) and (3) can be rearranged to yield

$$\Delta E = (E - E_0) = \tfrac{1}{2}(P + P_0)(V_0 - V) \qquad (4)$$

This equation is often referred to as Rankine-Hugoniot (R-H) relation.

The Rankine-Hugoniot equation of state (EOS) is a nearly adiabat compressibility curve in the P-V plane. If, however, the shock wave propagates through a laminated medium composed of materials with different shock impedances, it ceases to follow the Hugoniot adiabat but rather achieves isentropic behaviour with only a moderate temperature increase.

From (1) and (3), and using (4) one obtains

$$U_s^2 = V_0^2 \left(\frac{P}{V_0 - V}\right) \qquad (5)$$

$$U_p^2 = P(V_0 - V) \tag{6}$$

if $P_0 = 0$, $U_{p0} = 0$, and V and V_0 = specific volume behind and ahead of the shock front.

Any set of U_s's and U_p's correspond on the P-V diagram (Fig. 1) to a straight line U_s = const and a hyperbola U_p = const.

The intersection of the curves U_s = const and U_p = const fixes the state of the shock compression with parameters P_1, V_1 in A (see Fig. 1). The aggregate of the states which arise when the substance is compressed by shock waves of different intensity, determines the position of the shock-compression curve – the Hugoniot adiabat P_H passing through A, (Fig. 1)[22]. As can be seen from the Eqs. (5) and (6), measuring the quantities U_s and U_p enables the determination of the shocked state of the material, i.e., the calculation of the pressure, P_1, and the specific volume, V_1, attained by the material as a result of the shock. Numerous methods and techniques were developed and used successfully in the last two decades for measurements of shock and particle velocities, and wave profiles. This important and broad field is beyond the scope of the present article. The reader is referred to a critical review by Graham and Asay[23] on the measurement of wave profiles in shock-loaded solids.

The pressure P_1 attained by shock-compression of a solid, and resulting in the specific volume $V_1 < V_0$ is composed of two contributions P_c and P_t (Fig. 1). P_c is called the elastic or "cold" pressure, and results from the strong repulsive force of the interatomic potential. P_t is referred to as thermal pressure, and is associated with the thermal motion of atoms and electrons due to the shear compression. Therefore, as follows from Eq. (4), the total increase in internal energy (ΔE) is equal to the area of the triangle OAB in Fig. 1. This energy increment consists of an elastic component, ΔE_c (curvilinear triangle OCB) which is a result of elastic (cold) pressure developing in the solid as discussed

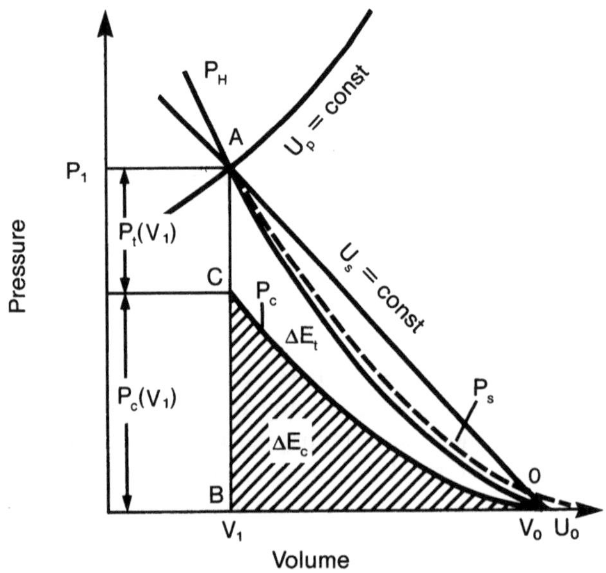

Fig. 1. Pressure-Volume relations in a solid material (after Alt'shuler[22]). P_c – Cold compression curve; P_H – Hugoniot adiabat; P_s – Expansion isentrope of shock-compressed material, P_t – thermal pressure; U_p – particle velocity; U_s – shock velocity; ΔE_c – change in internal energy due to cold compression; ΔE_t – change in internal energy due to thermal pressure

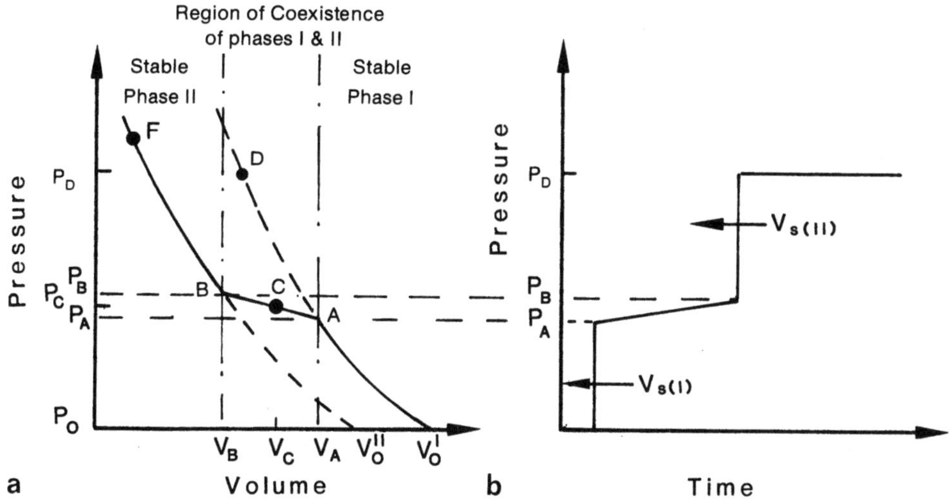

Fig. 2a, b. Shock adiabats (**a**) and pressure history (**b**) of a material undergoing a phase-transition. **a)** V_0^IAD – shock adiabat of the stable phase I, $V_0^{II}BF$ – shock adiabat of the stable phase II, **b)** $V_{s(I)}$ – shock velocity in phase I, $V_{s(II)}$ – shock velocity in phase II

above, and a thermal energy component, ΔE_t, represented by the curvilinear triangle OAC.

The shock transition provides both kinetic and internal energy to the material through which the shock wave propagates. Moreover, irreversible work is done on the material as it passes through the shock front. It can be shown[12)] that the entropy increases monotonically with pressure along the R-H shock adiabat, i.e., the R-H curve lies above the adiabat passing through the initial state.

Inasmuch as $\Delta E > \Delta E_c$, the shock-compression process is accompanied by heating of the substance and by an increase in its entropy, which in turn leads to the appearance of a thermal pressure component P_t. As can be seen from Fig. 1, the thermal energy and the thermal pressure increase progressively with increasing shock pressure.

For most materials, the shock-wave velocity, U_s increases with shock pressure. However, there are two important exceptions from such behaviour of a material under load. These are the elastic-plastic wave structure arising from material rigidity and the two-wave structure associated with certain phase changes in materials.

The splitting of the shock wave is related to the anomalous behaviour of the Hugoniot adiabat for the material in the region of a phase transition. The P-V diagram for a material undergoing a polymorphic transformation is shown schematically in Fig. 2a. When the material is compressed from its original volume V_0^1 to beyond that of a certain state with a volume of V_A, phase transition of phase I into phase II begins. The crystal lattice starts to rearrange itself in such a way that the new equilibrium positions of the atoms correspond to smaller interatomic distances[2]. Therefore, assuming that the specific

2 This, however, is not true for an anomalously behaving porous material[22)] which can decrease its density upon shock loading (under particular conditions) and for some solid (non-porous) materials, which also decrease their density upon shock loading. The loosening of the structure, and the saturation of the structure with defects creating a larger unit cell volume proceeds not only by a physical route (i.e., as a result of size reduction and misorientation of the crystallites) but also by a chemical one, such as reduction of the valences of the atoms, which lead to an increase in the unit cell parameters[24, 25)]. These two cases are excluded from the discussion

volume of the substance is decreasing during the phase transition, a much smaller increase in pressure will be required in the transition region AB (Fig. 2b) than in the original phase. If there was no rearrangement, the pressure curve would have extended upward along the R-H curve of the original phase (path AD in Fig. 2a). Complete rearrangement of the lattice and, thus complete transformation of the material to state II ends at point B. Beyond this point, the system follows the P-V curve of the new phase in a regular way, i.e., along path BF. The slopes of the curves in single phase regions differ from each other because the compressibilities of these phases are different.

It can be shown[26)] that a two-wave structure will arise if

$$\frac{P_A - P_0}{V_0^1 - V_A} > \frac{P_B - P_A}{V_A - V_B}, \tag{7}$$

where V_0^1, P_0, V_A, P_A, P_B are the conditions on the shock adiabat, and $P_0 < P_A < P_B$ and $V_0^1 > V_A > V_B$ (Fig. 2).

Figure 2b illustrates how a two-wave structure will arise in a material undergoing a phase transition. In phase I the shock wave propagates with velocity $V_{s(I)}$. In the transformation zone to which both phases contribute, the velocity of the shock will change gradually until the transformation is completed. Finally, the shock wave will acquire the velocity $V_{s(II)}$.

Shock waves propagate through polycrystalline or porous materials in a much more complex pattern than through single crystals. On the scale of each individual crystal grain, the shock waves at first bring about the same compression as in the case of a continuum, but as the shock wave begins to break through to the free surface of the grain, there is an abrupt change in pressure and the rarefaction wave accelerates the particle to twice the mass speed[27)]. As a result of multiple collisions between crystal grains, strong local pressures are generated which cause additional compression, heating, and crushing of the particles. The overall thermodynamic pattern of shock propagation through a powder is therefore highly nonuniform, showing micropockets of high pressure and temperature ('hot spots') next to regions of almost ambient pressure and temperature in the voids between the grains. Therefore, the compression of a porous material is generally less severe than the compression of a similar monolithic solid, at the same external pressure.

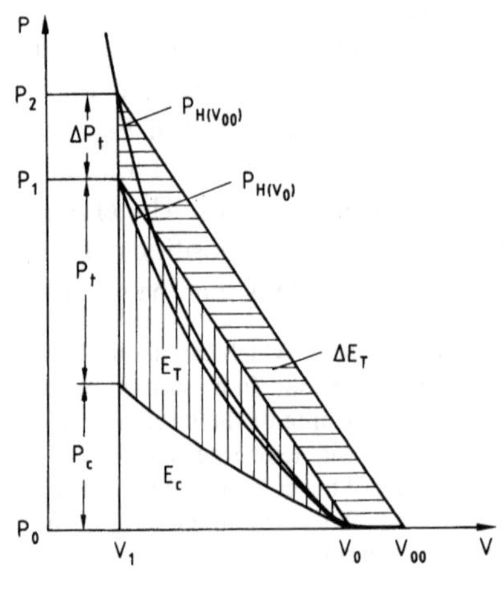

Fig. 3. P-V shock adiabats for solid *($P_H(V_0)$)* and porous *($P_H(V_{00})$)* materials. P_c – cold pressure, P_t – thermal pressure, E_c – cold (elastic) internal energy, E_T – internal energy due to the thermal pressure, ΔE_T – increase in internal energy due to the porous nature of the material, ΔP_t – excess thermal pressure due to the porous nature of the shocked material

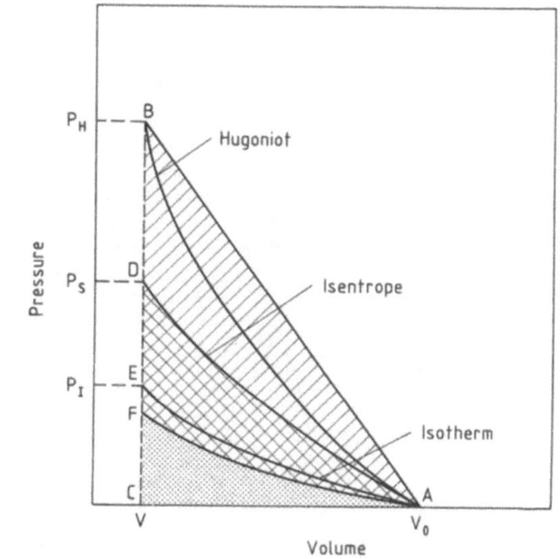

Fig. 4. P-V diagram illustrating the relations between Hugoniot, isentrope and isotherm

Figure 3 demonstrates how the compression of a porous sample with initial volume V_{00}, represented by the Hugoniot adiabat $P_H(V_{00})$, differs from compression of a solid sample with initial volume V_0. As can be seen from the figure, an additional increase in thermal energy ΔE_T is created in the porous sample which causes an additional increase in the thermal pressure component, ΔP_t. As a result the "porous" adiabat will always lie above the "solid" one.

Figure 4 explains the relations between Hugoniot, isentrope and isotherm curves in a P-V plane. Curves represented in Fig. 4 are projections onto the P-V plane from an equation-of-state surface in P-V-E or P-V-T space as shown in Fig. 5. For example, an isotherm in this plane is the projection onto the P-V plane, of the intersection of the T = const plane with the P-V-T equation-of-state surface.

The isentrope and isotherm represent a series of states which can be realized continuously. The Hugoniot represents the locus of all states which can be reached by shocking a material from a given initial state.

For these three thermodynamic processes the change in internal energy is

$$\Delta E = \int_{S_0}^{S} T dS_T - \int_{V_0}^{V} P dV_T = \int_{V_0}^{V} T \left(\frac{\partial P}{\partial T}\right)_v dV_T - \int_{V_0}^{V} P dV_T \quad (8)$$

for an isotherm,

$$\Delta E = - \int_{V_0}^{V} P dV_s \quad (9)$$

for an isentrope, and

$$\Delta E = \tfrac{1}{2}(P + P_0)(V_0 - V)$$

for a Hugoniot (see also Eq. 4).

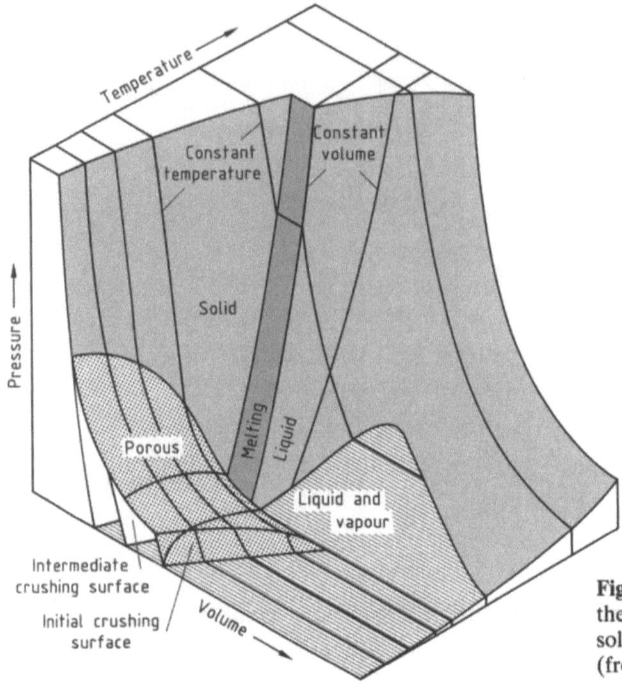

Fig. 5. Schematic representation of the P-V-T relationships for typical solid and porous material (from Isbell et al.[30])

As discussed earlier for a Hugoniot curve, the areas lying below the isentrope curve (AD in Fig. 4) and isotherm curve (AE in Fig. 4) represent the changes in internal energy given by Eqs. (8) and (9). Thus, the isentrope will always lie below the Hugoniot, since $(\partial P/\partial E)_V > 0$. It should be noted that the energy required to compress a sample isothermally is less than the area under the curve AE by the first term on the righthand side of Eq. (8) which is always negative. Therefore, it is represented by the area under the curve AF. The increase in energy from isotherm through isentrope to Hugoniot leads to progressively higher temperatures for a given compression.

The three-dimensional P-V-T space shown in Fig. 5 contains all possible Hugoniots equation of state (EOS) for non-porous and porous materials. Each point of the non-porous EOS surface represents a unique state derived under the assumption of Hess' theorem, i.e., the pressure at a particular specific volume and temperature is independent of the path used to arrive at that volume and temperature. On the other hand, the EOS for porous materials cannot be described by a single equation but by a mathematical model which considers the "crushing" history of the sample as described by a set of crushing surfaces.

Because of such anomalous behaviour in porous materials, it is very difficult and sometimes impossible to compress a porous material to its solid state (pore-free) even with a very strong shock wave. This problem limits the application of shock waves in fields of physics where a closer contact of atoms than in the parent body is desired. An obvious solution to this problem would be to create a shock wave in a substance in such a way as to separate the pressure pulse and the thermal pulse of the shock.

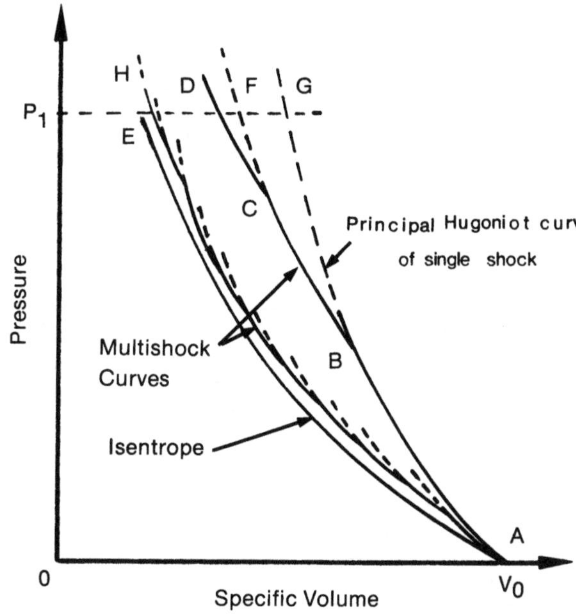

Fig. 6. Hugoniot curves for a solid under a single compression *(AG)* and multiple compression *(ABCD, AH)*. Note that the fan of weak shock waves nestles against the isentrope *AE*

If, instead of a single shock wave with amplitude P, a fan of weaker shock waves with increasing amplitudes strikes a substance, then the final and initial states of the substance will be connected not by a single shock adiabat, but by a series of adiabats which are created by the waves in the fan (Fig. 6). Each of these adiabats and, therefore, the total compression is very close to the isentrope[28]. A brief discussion of some of the methods which allow to break a single shock into a series of shocks can be found in Ref. 28.

Those interactions are very important in shock physics that occur when (i) two shock waves collide, (ii) overtake one another, or (iii) interact with an interface. Transitions between dynamic states may also occur through rarefactions, which reduce material to a lower pressure state and accelerate it in a direction opposite to that of propagation of the original shock. For a shock propagating into a material at initial state ϱ_0, P_0, U_{p0} and compressing it to a higher pressure state ϱ_1, P_1, U_{p1}, where ϱ is the density, P the pressure, and U_p the particle velocity, a graph of P vs. U is very instructive (Fig. 7). From

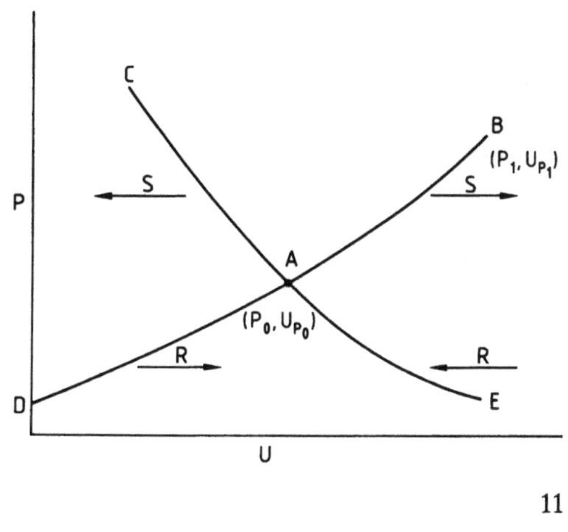

Fig. 7. Schematical presentation of wave-transitions in P-U plane[12]. *S* shock wave, *R* rarefaction wave

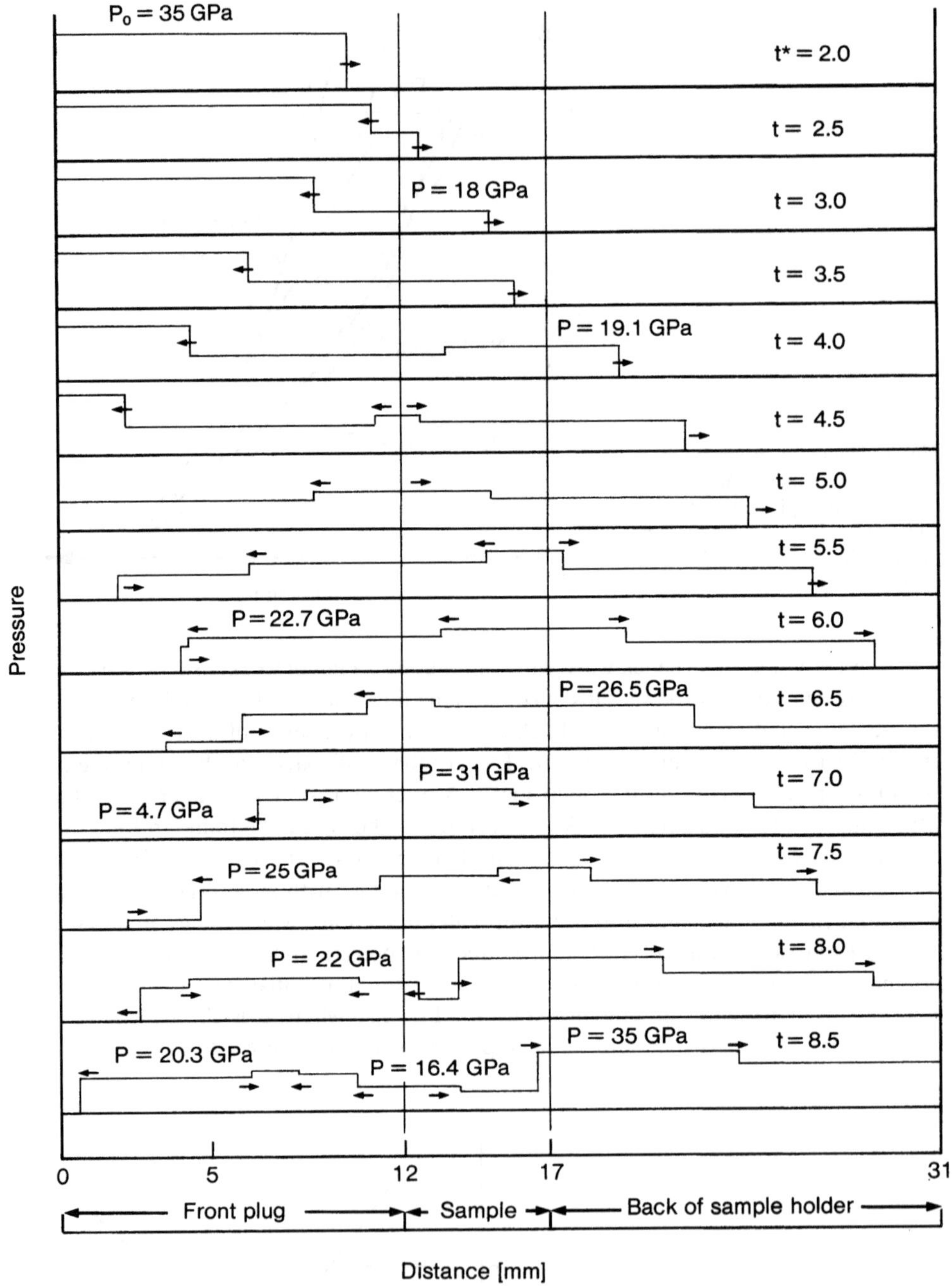

Fig. 8. Pressure history in a composite consisting of 3 different materials calculated for 14 consecutive time intervals.(*) Time is given in μs

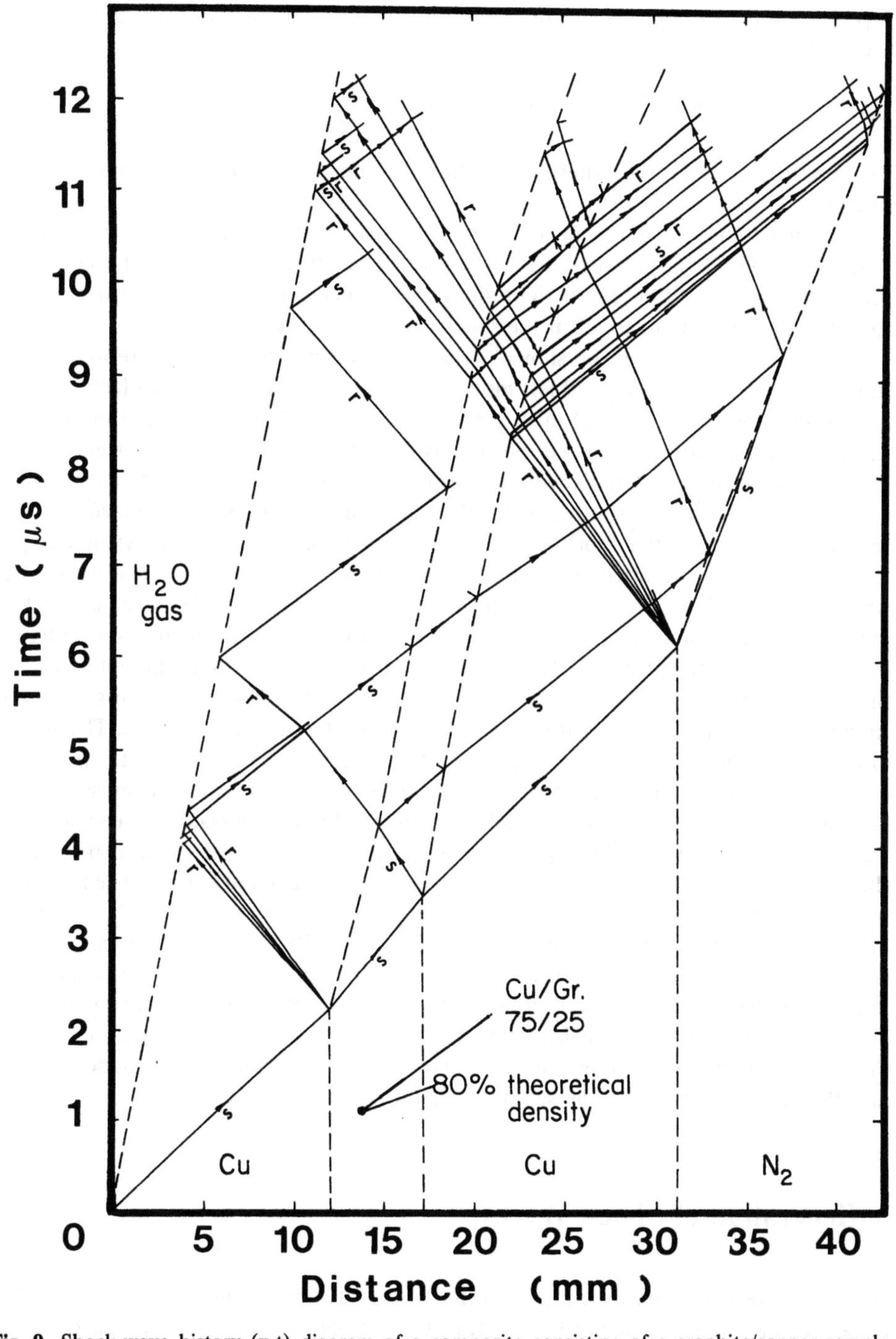

Fig. 9. Shock-wave history (x-t) diagram of a composite consisting of a graphite/copper sample sandwiched between two copper plugs, *s* – shock, *r* – rarefaction (from Kleiman et al.[5])

Eqs. (1) and (2) and superposing on the system a uniform particle velocity U_{p0}, it can be shown that

$$U_{p1} - U_{p0} = \pm \sqrt{(P_1 - P_0)(V_0 - V_1)} \, . \tag{10}$$

Equation 10 represents the hyperbolic relation (6) when a uniform flow is imposed on the system.

The two solutions of Eq. (10) are drawn as curves AB and AC in Fig. 7. Shock transitions from P_0, U_{p0} to any other state must lie on either AB or AC. The shock will accelerate material from left to right along AB, which represents the positive sign in Eq. (10), and from right to left if the shock transition is laying on AC.

In a similar way, it can be shown[12)] that the branches AD and AE represent two solutions for a right moving and left moving rarefaction wave. Thus a rarefaction propagating into the state (P_0, U_{p0}) must produce a final state laying on AD if the wave is travelling to the right and on AE if it is travelling to the left. If a shock wave travels through two or more materials with different specific densities in intimate contact with each other, it will be transmitted and reflected in different ways at the interfaces. The relative magnitude of the shock and rarefaction waves depends on the dynamic impedances of the material at the interface. The dynamic impedance is defined as $\varrho_0 \cdot U_s$, where ϱ_0 is the initial specific density of the material and U_s is the shock velocity in this material. Figure 8 illustrates a one-dimensional pressure history in a composite structure consisting of 3 different materials and calculated for 14 consecutive time intervals. At time $t = 0$ (not shown) a shock with an amplitude of 35 GPa enters the first material from the left and propagates through it until it strikes the first interface, etc. The arrows in Fig. 8 show the direction of the shock front propagation. By combining the pressures calculated or measured in the materials with calculated temperatures accompanying the high pressures, a dynamic diagram can be constructed for a given material (see Chap. 4). Figure 9 gives another useful representation of a shock-wave history in a specimen[5)], showing the positions of waves and material surfaces as a function of time. The front and the back of the copper plugs are bounded by gas. The initial shock wave enters the front of the copper plug at time zero with an amplitude of 50 GPa. At the front of the graphite/copper sample it is reflected as a rarefaction wave (r-r) and retransmitted as a shock (s) into the sample. Subsequent collisions of the waves with the interfaces are shown. The dashed lines represent the moving contact surfaces. It can be seen that the thickness of the sample diminishes greatly. Since permanent phase transition to diamond was not modeled in this calculation, the sample is seen to expand again.

3 Generation of Shock Waves

A shock wave is created by a rapid acceleration of a surface caused by a very sudden release of mechanical, chemical, electrical, nuclear, or radiation energy in a confined space. Energy densities in shock waves are very high, and their generation will depend on the existence of energy sources, that must deliver energies in the range of megajoules in a time of the order of microseconds. Table 2 illustrates some of the most important primary energy sources available on earth.

Table 2. Terrestrial energy sources for the generation of high energy density (after Knoepfel[47])

Primary source	Final form	Energy density ($MJ \cdot cm^{-2}$)	Total involved energy (MJ)
Chemical explosives	–	$8 \cdot 10^{-3}$	100
	Metallic jets	1.0	10^{-3}
	Flyer plates & projectiles	0.8	3.0
	Direct contact explosives	–	–
Energy stored in capacitors	–	10^{-8}	5
	Exploding wire	0.05	10^{-3}
	Magnetic force accompanying the discharge through a capacitor	–	–
Optical storage	–	10^{-6}	10^{-3}
	Focused laser beam	0.4	10^{-5}
Nuclear explosion	–	10^{4}	10^{11}

The motion of the shock wave through the specimen creates compression states and expansion states. The compression states consist of a fast, smooth pressure rise, in which the compression is approximately isentropic, or involves a shock compression, in which the pressure rise is extremely fast and in which irreversible heating occurs, i.e., the entropy of the system increases. The after-shock expansion state usually involves a rapid isentropic expansion and not a shock.

Different techniques are being used to produce shock waves in solid materials. The final goal of the experiment usually dictates the type of loading conditions. If the goal is the study of material properties, in response to shock loading, then well-defined and controlled conditions must be created and, therefore, a controlled planar shock wave is desirable. If the final goal, however, is material synthesis or post-shock recovery studies shock loadings other than planar ones can be used, i.e., cylindrical, hemispherical or spherical.

A description of the basic physical principles underlying those shock-wave configurations can be found in[29]. Most of the techniques used for shock-wave creation are exhaustively described in the literature[23]. Here only a short review of the methods of generation of shock waves relevant for synthesis of superhard material will be given. More detailed descriptions of devices used to produce superhard materials are presented in Chap. 5.

3.1 Direct Contact High Explosive Methods

Shock waves of planar, cylindrical or spherical structure can be created in solid materials using a high explosive charge in direct contact with the sample or with a preservation plate or ampoule. In this case the detonation pressure of the explosive, P_{det}, should be equal to or exceeding the pressure immediately behind the shock front, P_s. The total amount of energy exchanged by the shocked material during compression and subsequent isentropic expansion is[31]

$$E = E_2 - E_0 + \tfrac{1}{2} U_p^2 , \tag{11}$$

where the subscript 0 refers to the initial state of the material, subscript 2 refers to the state attained by the material in isentropic expansion and U_p is the particle velocity in state 2.

Using the conservation of momentum Eq. (2) and Eqs. (5) and (6) it can be shown[31] that

$$P_s = \varrho_0 U_s^2 \frac{\Delta V}{V_0} \tag{12}$$

and

$$P_{det} = \varrho_{det} U_s^2/4 \tag{13}$$

where ϱ_0 and ϱ_{det} are the initial densities of the specimen and of the explosive respectively, U_s, is the velocity of the shock wave, V_0 is the initial specific volume of the specimen, and ΔV is the change of the specific volume of the specimen resulting from compression.

From the requirement $P_{det} \geqq P_s$ it follows that

$$\frac{\varrho_{det}}{4\,\varrho_0} \geqq \frac{\Delta V}{V_0} . \tag{14}$$

This inequality provides a convenient estimate of the feasibility to employ the contact method of shocking to achieve a desired compression. Since the densities of commonly available explosives are usually less than the initial density of the specimen[31], the inequality (14) is satisfied only when the relative compression $\frac{\Delta V}{V_0}$ is small, thus making the direct contact method suitable only for shocking materials with nearly theoretical density up to moderate pressures.

Explosive lenses which usually consist of a combination of fast-detonating and slow-detonating explosives (Fig. 10) with diameters of 100 to 300 mm proved to be good sources of planar shock waves[31–34]. The geometry of the two-explosive lens shown in Fig. 10 is chosen so that the high-velocity detonation wave travelling down the outside of the lens ($U_{det.1}$) initiates the detonation of the inner explosive ($U_{det.2}$). The angle α is chosen so[12] that

$$U_{det.2} = U_{det.1} \cdot \sin\alpha . \tag{15}$$

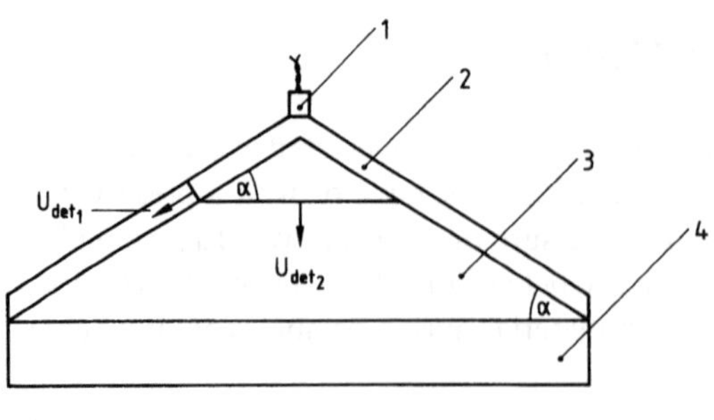

Fig. 10. Direct Contact High Explosive Method Using a Two-Explosive Lens. *1* – Detonator; *2* – High Velocity Explosive; *3* – Low Velocity Explosive; *4* – Specimen

Instead of a two-explosive lens a different two-stage device (called the "Mousetrap") is often used to initiate a planar wave in a specimen attached to the high explosive (Fig. 11). A pad of high-velocity explosive (1) in contact with a driver plate (2) is ignited at one end by a detonator (3). The detonation wave propagates down the pad with velocity, U_{det}, and the high pressure of the detonation gases accelerates the driver plate to velocity, V. If the driving explosive (1) with the attached plate (2) are inclined at an angle $\alpha = \sin^{-1} (V/V_{det})$ to the explosive surface (4) to be detonated, the plate (2) will strike every point of surface (4) simultaneously and, if its velocity is high enough, will initiate a planar detonation wave which, in turn, will accelerate the driver plate (5). If angle α satisfies the condition discussed above, the third pad of high-explosive (6) will be stricken simultaneously by the driver plate (5) and a planar shock-wave will be transmitted into the sample (7).

Detonation of an explosive over a large area of the sample generates a free compression wave travelling with a velocity determined by the sample material and the pressure generated. This compression, even if applied not instantaneously over the whole area, builds up rapidly to a shock compression wave with a steep front which by its nature is a discontinuity in pressure, density, and internal energy. Pressures in the range 4–40 GPa[23)] were achieved with such planar lenses. By enclosing the sample material in an ampoule of cylindrical or spherical configuration and by suitably shaping the high explosive around the ampoule, cylindrical or spherical shock waves can be created in the material[31, 33, 35)].

In the case of cylindrical shocks, the high pressure is applied progressively to a cylindrically shaped sample by detonating a layer of explosive surrounding the sample. The detonation front then moves in the direction of the samples larger linear dimension. When the length of the cylinder greatly exceeds its diameter, the pressure wave attains practically steady-state conditions over much of the cylinder[31)].

3.2 Explosively Driven Flyer Plates and Projectiles

Flyer plates with the planar lenses described in Chap. 3.1 can be used if pressures are required higher than those attained by direct contact high explosives. A relatively thin flyer plate is accelerated by a plane-wave explosive system. After a free run of

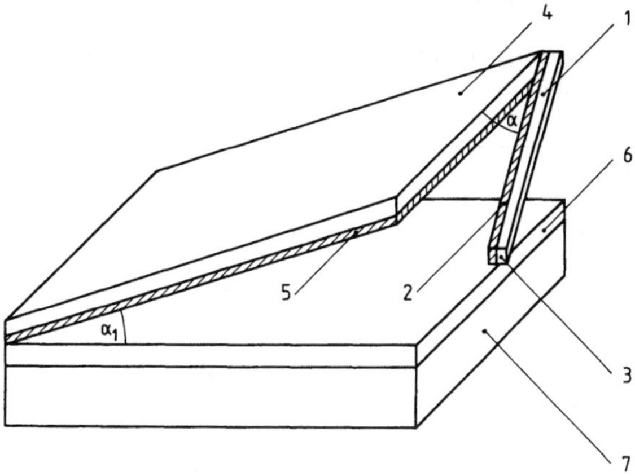

Fig. 11. Direct Contact High Explosive Method Using a "Mousetrap" Lens. *1,4,6* – Explosives; *2,5* – Driver Plates; *3* – Detonator; *7* – Specimen

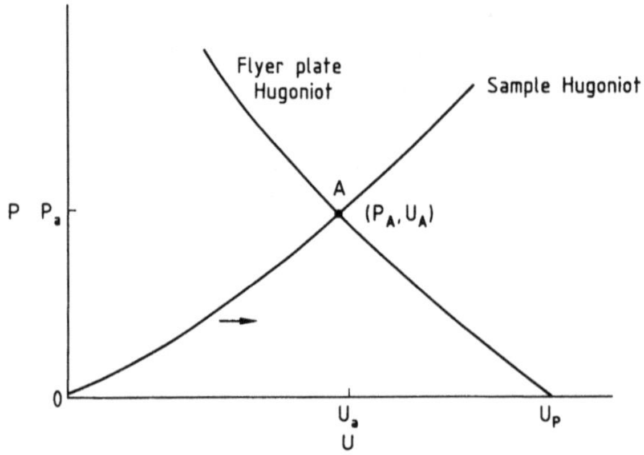

Fig. 12. Schematic explanation in P-U plane of creation of a planar shock wave induced by impact of a flyer plate

20–40 mm, this plate collides with the target plate which usually constitute the sample or to which the sample is attached.

The detonation wave produced by the plane-wave explosive system accelerates the flyer plate to a high velocity U_p during the free run in the gap between the flyer and the sample. Impact produces a right-travelling wave in the sample, which is initially in state $p = 0$, $u = 0$, (Fig. 12). The shock induced in the sample lies on the direct R-H curve of the sample, the stopping shock created in the flyer plate lies on the reflected R-H curve through $U = U_p$. The common state produced by impact is described by P_A, U_A which is the intersection of the R-H curves of the sample and the flyer. Depending on the thickness of the flyer plate, and the amount of explosive, pressures as high as 100 GPa can be generated with this configuration[26]. The duration of the high pressure pulse is dependent upon the flyer plate thickness. Table 3 illustrates how the thickness of the flyer plate affects the particle size distribution in explosive synthesis of diamonds. In both cases shown in Table 3 the pressure upon impact was 72 GPa[36]. If the flyer plate is made from a dense high-impedance material such as steel, the impact can generate considerably higher pressure than the explosive itself in direct contact methods. For this reason the flyer-plate technique has a much wider range of applications since the pressures generated can be adjusted by the shock-impedance properties of the flyer and the target, respectively.

Table 3. Size distribution of diamond particles in two experiments with different flyer plate thicknesses (from Andreev et al.[36])

Size distribution of formed diamonds (μm)	Percentage of each size-fraction in the experiment (%)	
	Flyer plate 4 mm thick	Flyer plate 1.5 mm thick
60–40	0.6	–
40–28	1.0	–
28–20	2.0	0.3
20–14	4.6	2.2
14–10	5.2	1.4
10– 7	4.6	4.4
7– 5	5.0	4.8
5– 3	7.0	11.7
3– 2	10.7	15.9
2– 0	59.6	59.3

In the projectile impact loading technique, the ranges of impact pressures, and the duration of the pressure pulse can be varied on a practically continuous scale. The loading conditions can be defined more precisely than in other techniques because the projectile velocities can be measured to an accuracy of $\pm 0.1\%$[23]. The use of gas gun drivers makes this technique relatively safe and adaptable to the laboratory environment. All these advantages of the projectile impact technique renders it very flexible and versatile.

Figure 13 shows schematically a planar projectile impact experiment arrangement. This technique involves the use of a driver (1) to accelerate the projectile (2) upon the rupture of a diaphragm (3). The projectile after being accelerated along the barrel (4) impacts the sample (5). Depending on the type of driver used, the length of the barrel can vary from 3 to 25 m. Drivers utilizing compressed gas[37, 38], propellants[39] or two-stage gas guns[19, 40] are used to accelerate the projectile. The diameter of the bore of the gun that dictates the projectile diameter can vary from 25–30 mm in two-stage gas guns[19, 40] and up to 50–100 mm in compressed gas drivers[37]. Misalignment or tilt between projectile and specimen surfaces can affect the maximum pressure attained upon collision. Special care, therefore, has to be taken considering mechanical alignment and wobbling[19, 30].

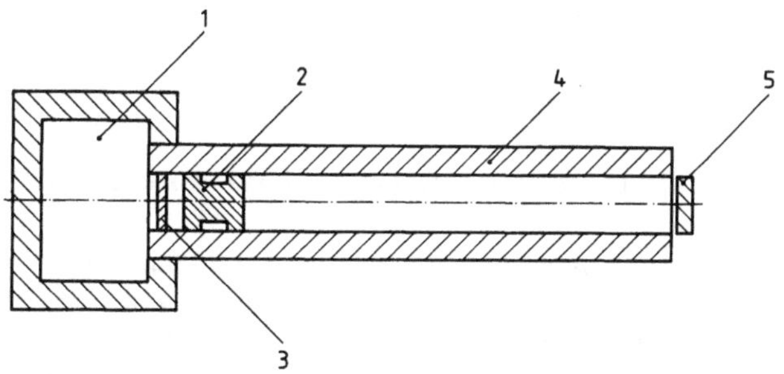

Fig. 13. Projectile Acceleration Method (Keeler and Royce[41]). *1* – Driver; *2* – Projectile; *3* – Diaphragm; *4* – Barrel; *5* – Sample

3.3 Pulsed Radiation

The duration of a radiation pulse can be made very short compared to the time it takes for an acoustic wave to traverse the irradiated material. Therefore, in most cases, the energy deposition can be considered to be instantaneous.

Using powerful electron beams, pulsed lasers and pulsed neutron beams, pressures varying from a few GPa up to 2 TPa[42] can be achieved.

The variability of the laser flux density that can be obtained by varying the laser power, and focusing conditions can be used to change the thermodynamical state of different materials over a wide range.

Pulsed lasers developed for inertial confinement fusion applications can produce over 20 TW peak power that can be focused onto an area of less than 10^{-4} cm^2, giving a power flux greater than 10^{21} W/m^2 [43]. At power densities of only 10^{17} W/m^2, the laser beam vaporizes and ionizes any solid surface. The radiation pressure associated with absorption or reflection of such a flux (10^{21} W/m^2) is of the order of 10^{13} Pa (10 TPa). Pressures as high as 10^{12}–10^{14} Pa have been experimentally generated using such lasers[43].

In the early studies of the effects of laser light absorption by solid surfaces, a simple relationship between the absorbed light intensity, I, and the ablation-induced pressure, P, was derived[44, 45], yielding

$$P = \frac{8}{5} \cdot \left(\frac{\varrho_c}{4}\right)^{1/3} \cdot I^{2/3}, \tag{16}$$

where ϱ_c is the plasma density at a critical surface site where the local electron plasma frequency matches the laser frequency of the normally incident radiation pulse. The stress distribution associated with the deposited energy along the axis of stress propagation is

$$\sigma_x(X) = \frac{\gamma_x}{V} E(x) \tag{17}$$

where σ_x is the stress component, E(x) is the absorbed energy per unit mass, γ_x is the Grüneisen parameter in the direction of radiation propagation and V is the volume. Numerous studies on pulsed radiation and other laser applications were performed over the last 15 years. A short review of them can be found in[23, 46].

3.4 Capacitors as Sources of High Energy Density

Magnetic forces accompanying large current pulses from the discharge of capacitor banks have been utilized to generate strong shock waves. If the magnetic field in such experiments is of the order of GA/m and its rise time is in the microsecond range, a pressure pulse may eventually build up in front and form a shock wave which can be characterized by parameters like U_s and U_p.

The magnetic pressure[47)]

$$P_H = \frac{\mu_0 \mathfrak{H}^2}{2}$$

($\mu_0 = 4\,\pi \cdot 10^{-7}$ H/m, $\mathfrak{H}$ = magnetic field strength) drives a shock wave into the conductor around which the field is built. It can be shown[47] that by equating the magnetic pressure to the Hugoniot pressure given by the conservation of momentum relation (Eq. (2), a useful relationship between the shock parameters (U_s, U_p) and the applied magnetic field strength can be derived as follows:

$$\varrho_0 \cdot U_s \cdot U_p = \frac{\mu_0 \mathfrak{H}^2}{2}, \tag{18}$$

where ϱ_0 is the initial density of the material the conductor is built from, and U_s and U_p are shock and particle velocities in the conductor, respectively.

It can be also shown that the pressure induced in the conductor is related to the current, I, that induces the magnetic field, $\mathfrak{H}$, as follows[48]:

$$P = \frac{\mu_0 I^2}{8\,\pi^2 r^2}\,, \tag{19}$$

where r is the initial radius of the conductor in which the current, I, is flowing.

Another use of energy stored in a capacitor is made in exploding-wire experiments. In these arrangements a long thin conducting cylinder is impulsively heated by a high-energy capacitor discharge. Depending on the choice of the conductor, the initial conditions, and the type of housing accommodating the exploding conductor, a wide range of shock pressures can be generated. The theory describing the processes of exploding wires is quite complicated[49)] and should not be discussed here.

Some practical applications of the methods described in this section will be further discussed in Chap. 5.

4 Phase Transitions in Carbon and Boron Nitride

4.1 Geometry of Direct Phase Transition

Rearranging the crystal structures of the starting materials, e.g., hexagonal graphite and g-BN required for shock-induced crystal growth of diamond, and w-BN and z-BN, respectively, involves movement of atoms out of the planar hexagonal rings of carbon atoms, and boron and nitrogen atoms, respectively. This out-of-plane movement of atoms results in either "buckling" of the planar hexagons to yield the boat-like configura-

Table 4. Crystallographic data of phases of carbon (A) and boron nitride (B)

Phase	Structure type	a_0 (nm)	c_0 (nm)	Density (Mg/m^3)	Reference
(A)					
Graphite	D^4_{6h}-$P6_3/mmc$	0.246	0.669	2.284	50
Rhombohedral	D^5_{3d}-$R\bar{3}m$	0.246	1.006	–	51
Graphite (r-C)	-R3	0.245(6)	1.004(4)	2.280	52
Lonsdaleite	D^4_{6h}-$P6_3/mmc$	0.252	0.412	3.51	53, 54
Diamond	O^7_h-Fd3m[a]	0.356(6)	–	3.514[b]	52
(B)					
Graphitic BN	D^4_{6h}-$P6_3/mmc$	0.250(4)	0.665(6)	2.34	58
Rhombohedral BN (r-BN)	$R\bar{3}m$	0.250(4)	1.001	2.29	59, 60
Hexagonal BN (w-BN)	$P6_3mc$	0.255(3)	0.422(8)[c]	3.454	9, 62
Cubic BN (z-BN)	$F\bar{4}3m$	0.361(6)	–	3.4870	9, 57
E-BN[d]	f.c.c.	0.8405?	–	2.55?	63

[a] According to Donnay[55)], the symmetry of natural diamond is $\bar{4}3m$ (see also[56)])
[b] Densities of types I (containing nitrogen platelets) and II are slightly different[57)]
[c] Unit cell length doubled according to[61)]
[d] Existence of this new high-pressure form not irrefutably established

tion characteristic for the wurtzite-type structures of carbon (lonsdaleite) and boron nitride (w-BN), or "puckering" to yield the chair-like configuration found in the zinc-blende-type structures (diamond and z-BN). In all cases an increase in density occurs (see Table 4).

Figure 14 shows the crystal structures of the hexagonal sheet-type (graphite, g-BN), rhombohedral sheet-type (r-graphite, r-BN), wurtzite-type (lonsdaleite, w-BN), and zincblende-type (diamond, z-BN) lattice arrangements. The crystallographic data as well as the density of various polymorphic modifications of carbon and boron nitride are shown in Table 4. The macroscopic orientation relations between principal lattice directions and planes are shown in Table 5.

4.1.1 Hexagonal Sheet-Type to Zincblende-Type Transformation

The C–C bond lengths in $\langle 11\bar{2}0\rangle$ direction of graphite differs from the C–C bond lengths in $\langle 110\rangle$ direction of diamond by only 2%. Transformation of graphite to diamond and g-BN to z-BN, respectively, involves movement of atoms out of the $\{0001\}$ plane of the graphite-structure to generate chair-like atomic arrangement ("puckering") of the hexagons as indicated in Fig. 14(d). Thus, a $\{0001\}$ plane of the graphite-structure becomes a diamond $\{111\}$ plane. Because there are three equivalent $\langle 11\bar{2}0\rangle$ directions in the graphite basal plane, three possible orientations exist for a[1)] diamond crystal. Shock-generated diamonds produced from 2H-graphite in fact show frequently a three-fold [112] texture[65–67)]. The possibility of diffusionless transformation of graphite into diamond via formation of an intermediate rhombohedral graphite structure was first proposed by Lonsdale et al.[17)] and experimentally observed by Vereshchagin et al.[68)], (see also 4.1.4). This transformation essentially requires only the out-of-plane shifting of atom by 0.25 Å to produce the puckered chair-like hexagon.

Table 5. Macroscopic orientation relations of parent and daughter structures in high-pressure transformations of carbon and boron nitride

(A)	Hexagonal sheet-type (G) → Zincblende (Z)	
	$(0001)_G \perp (111)_Z$ $(10\bar{1}0)_G \parallel (111)_Z$ $\langle 11\bar{2}0\rangle_G \parallel \langle 110\rangle_Z$ $\langle 0001\rangle_G \parallel \langle 112\rangle_Z$	
(B)	Wurtzite (W) → Zincblende (Z)	
	$(0001)_W \parallel (111)_Z$ $\langle 11\bar{2}0\rangle_W \parallel \langle 110\rangle_Z$ $\langle 0001\rangle_W \parallel \langle 111\rangle_Z$	
(C)	Hexagonal sheet-type (G) → Wurtzite (W)	
	Carbon $(0001)_G \parallel (10\bar{1}0)_W$ $\langle 11\bar{2}0\rangle_G \parallel \langle 11\bar{2}0\rangle_W$ $\langle 10\bar{1}0\rangle_G \parallel \langle 0001\rangle_W$	*Boron nitride* $(0001)_G \parallel (0001)_W$ $\langle 11\bar{2}0\rangle_G \parallel \langle 11\bar{2}0\rangle_W$

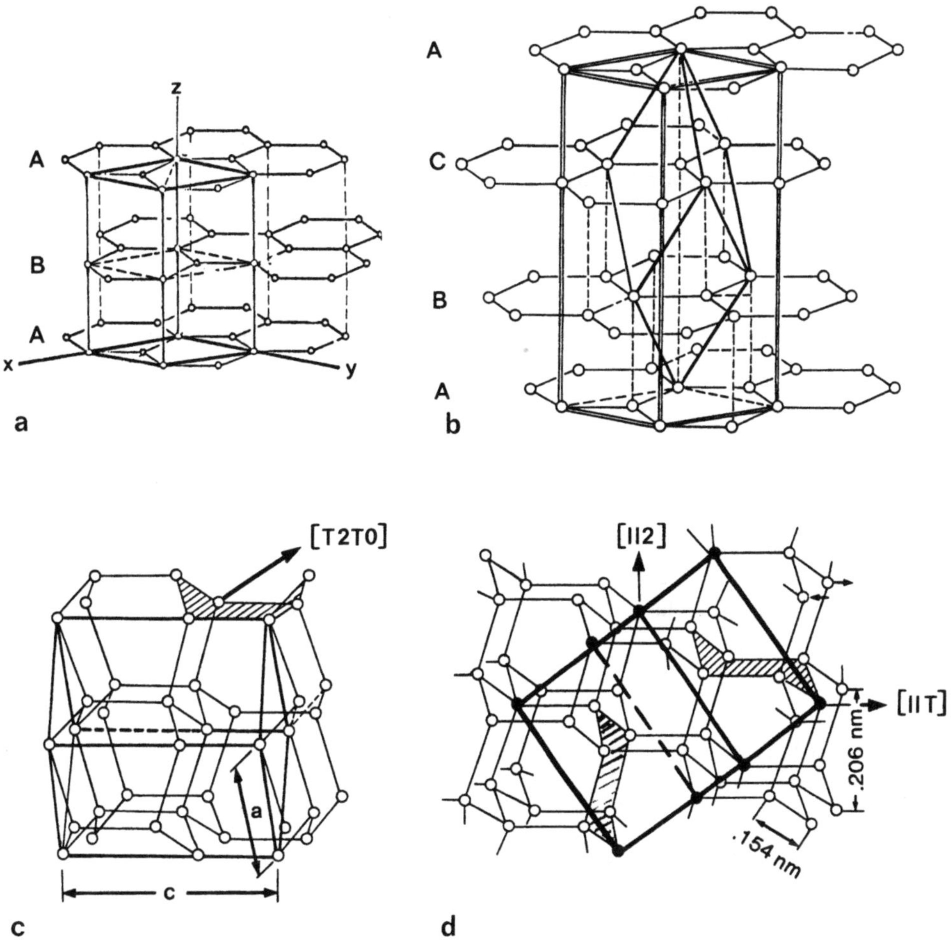

Fig. 14a–d. Crystal structures of **(a)** hexagonal sheet-type (graphite, g-BN), **(b)** rhombohedral sheet-type (r – graphite, r-BN), **(c)** wurtzite-type (lonsdaleite, w-BN) and **(d)** zinc blende-type (diamond, z-BN), polymorphic modifications of carbon and boron nitride (**a, b** after[64]; **c, d** after[65]). The outlines of the unit cells are given by the *heavy lines*

Transformation of g-BN into z-BN occurs by essentially the same mechanism. This has been confirmed by Sato et al.[69] who obtained z-BN by explosive shock compression of rhombohedral boron nitride (r-BN).

The activation energy of this phase transformation, however, should be considerably higher than that of graphite-diamond transformation since formation of the intermediate rhombohedral structure of boron nitride requires rotation of successive layers (Fig. 14) of the graphite-type boron nitride structure by 60°[70, 71]. This rotation is energetically favoured with increasing temperature owing to an increase in atomic mobility. Thus, z-BN is formed at high pressures predominantly at temperatures above 2700 K whereas at lower temperatures w-BN is formed whose change in transitional enthalpy, ΔH, is only 136 J/g[9].

4.1.2 Wurtzite-Type to Zincblende-Type Transformation

Transformation of the wurtzite (lonsdaleite, w-BN) into the zincblende (diamond, z-BN) structure involves a diffusionless displacive mechanism that is presumably activated by the passage of a train of Shockley partial dislocations[65] through the crystal lattice. As a result of this transformation, close-packed {0001} planes of the wurtzite structure become likewise close-packed {111} planes of the zincblende structure normal to the original graphite basal planes (Fig. 14, Table 5).

4.1.3 Hexagonal Sheet-Type to Wurtzite-Type Transformation

4.1.3.1 Carbon

Transformation of (hexagonal) graphite to lonsdaleite requires a cooperative slippage of reconstructive type with comparatively little change in free energy since out-of-plane "buckling" of the planar hexagons of the graphite to generate the boat-like atomic arrangement typical for the wurtzite structure is energetically favoured over "puckering" required to readjust the planar hexagons to the chair-like arrangement of the zincblende

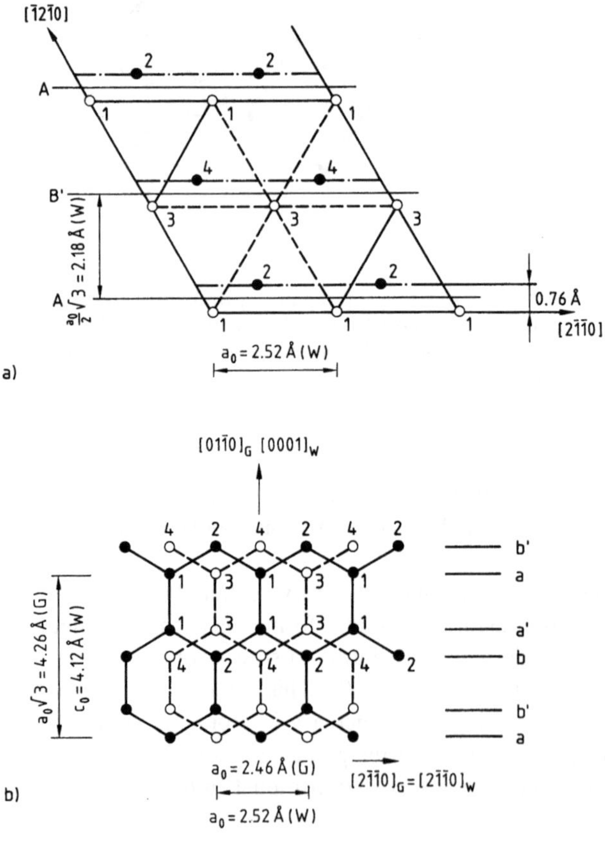

Fig. 15a, b. Transformation of hexagonal graphite into lonsdaleite (according to[73]), **(a)** Projection onto (0001) of intermediate structure of graphite (*A* and *B'*), and onto ($10\bar{1}0$) of lonsdaleite *(points 1,2,3,4)*, **(b)** Projection onto (0001) of lonsdaleite

(diamond) structure[72]. The {0001} plane of the graphite-structure becomes the $\{10\bar{1}0\}$ plane of the wurtzite (lonsdaleite)-structure, and the relationships hold $\langle 11\bar{2}0\rangle_G \parallel \langle 11\bar{2}0\rangle_w$ and $\langle 10\bar{1}0\rangle_G \parallel \langle 0001\rangle_w$ (Table 5). Figure 15 shows the crystallographic mechanism of the transformation. In Fig. 15(a) the shifts of the carbon atoms 1 and 2 in layer A, and 3 and 4 in layer B′ in opposite directions are shown, creating the boat-like arrangement of carbon atoms. Thus, an intermediate graphite-like structure AB′AB... is formed that requires shifting of layer B into position B′ by $(1/6)a[10\bar{1}0] = 0.71$ Å[3] of the original hexagonal graphite structure. This vector equals one-half of the Burgers vector of a partial Shockley dislocation in the graphite basal plane (see 4.1.4).

It should be emphasized that lonsdaleite retains the layered arrangement of the hexagonal sheet-type structure of graphite. Thus, there is only two-dimensional fit between the parent and daughter structures. An important consequence of this constraint is the need of diffusion for further growth of lonsdaleite to occur[53].

Compression of single crystals of graphite in c-direction leads to a drastic increase in electrical resistance at pressures between 14 and 15 GPa. Since this phase transformation to lonsdaleite is reversible upon pressure relieve at room temperature, it is thought to occur by solid-to-solid transformation without involvement of a liquid intermediate phase. If, however, the sample is annealed above 1270 K while pressurized, the transformation becomes permanent, and the lonsdaleite phase is retained. Although the thermal treatment apparently is not severe enough to cause melting, it promotes diffusion and in turn stabilization of the highly-strained lonsdaleite domains in the hexagonal graphite[74].

4.1.3.2 Boron Nitride

Transformation of graphite-like boron nitride into a w-BN can be achieved in a diffusionless way by a martensitic transformation mechanism involving splitting of the planar hexagons and thus generating planes occupied solely by boron atoms and nitrogen atoms, respectively, (Fig. 16(a)). In the hexagonal structure ABAB... of g-BN, layer A splits into two layers a′ and b, and layer B splits into layers b′ and a. Thus, the hexagonal structure of w-BN consists of the layer sequence ab′ba′ab′...

As seen in Fig. 16(b), the orientation relation between parent g-BN and daughter w-BN is

$$(0001)_G \parallel (0001)_w, \text{ and}$$

$$\langle 2\bar{1}\bar{1}0\rangle_G \parallel \langle 2\bar{1}\bar{1}0\rangle_w$$

Clearly no intermediate structure is formed as in the graphite-lonsdaleite transformation described in 4.1.3.1. This means that the transformation g-BN → w-BN could be achieved, in principle, athermally by simple lattice deformation without preliminary nucleation[75]. The pressure at which phase transformation may commence by an athermal martensitic mechanism has been calculated by Kurdyumov[75], using Estrin's concept[76], to be approximately 14 GPa. This value is in excellent agreement with experi-

3 The actual values given, respectively, by Kurdyumov[73] and Bundy and Kasper[53] are 0.76 and 0.73 Å

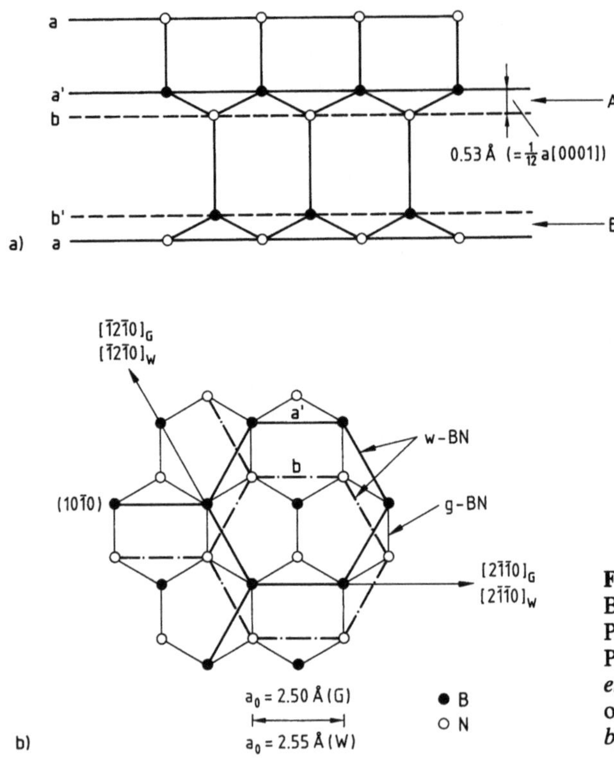

Fig. 16a, b. Transformation of g-BN into w-BN (according to[75)]). **(a)** Projection onto (10$\bar{1}$0) of w-BN; **(b)** Projection onto (0001) of g-BN *(layer A), (thin lines),* and projection onto (0001) of w-BN *(layers a' and b), (heavy lines)*

mental data obtained by static methods by Bundy and Wentorf[77)], and by shock-compression by Adadurov et al.[6)] and Coleburn and Forbes[9)]. The threshold pressure for the transformation g-BN → w-BN decreases with increasing temperature, giving finally way to thermally activated nucleation of the zincblende-type modification of boron nitride[78)], (see also 4.1.1).

4.1.4 Hexagonal Sheet-Type to Rhombohedral Sheet-Type Transformation

Contrary to the transformation of graphite-type structures of carbon and boron nitride into wurtzite- and zincblende-type structures described above, conversion of the hexagonal structure of graphite into the rhombohedral structure does not require elevated temperatures and high hydrostatic pressures.

It appears to be convenient to describe the structural differences of the two sheet-type modifications in terms of a hexagonal close-packed structure with a stacking sequence ABAB... and a face-centered cubic close-packed structure with a stacking sequence ABCABC... However, ideal h.c.p. and c.c.p. stacking sequences found in metals differ from those observed here, because the former show equal interatomic distances between co-planar and inter-planar nearest neighbours. In the sheet-type carbon structures, however, the interatomic distances in the basal plane are 0.141(5) nm but the interplanar distances are 0.334(5) nm for well-ordered graphite. Disordered, turbostratic graphite has a slightly larger layer distance of 0.344(0) nm[64)].

Formation of rhombohedral graphite was observed to occur by unidirectional (non-hydrostatic) pressure associated with shear[79–82]. X-ray single crystal studies[81] showed that unidirectional pressure alone does not suffice to produce rhombohedral graphite nor does hydrostatic pressure up to 1000 MPa. If a shear component is added, for example by prolonged grinding of graphite[79, 83], transformation gliding is induced[4]. The amount of rhombohedral graphite produced is proportional to the degree of shearing (= amount of gliding/thickness of crystals)[81]. A microscopic approach to the transformation mechanism has been developed by Boehm and Hofmann[80] and Laves and Baskin[81], and was shown to be closely related to the process of mechanical twinning involving relative movements of fixed carbon double nets in the direction of the carbon-carbon bonds by $a \cdot \sqrt{3}/3 = 0.141(5)$ mm. This idea was further developed by Freise and Kelly[82] who

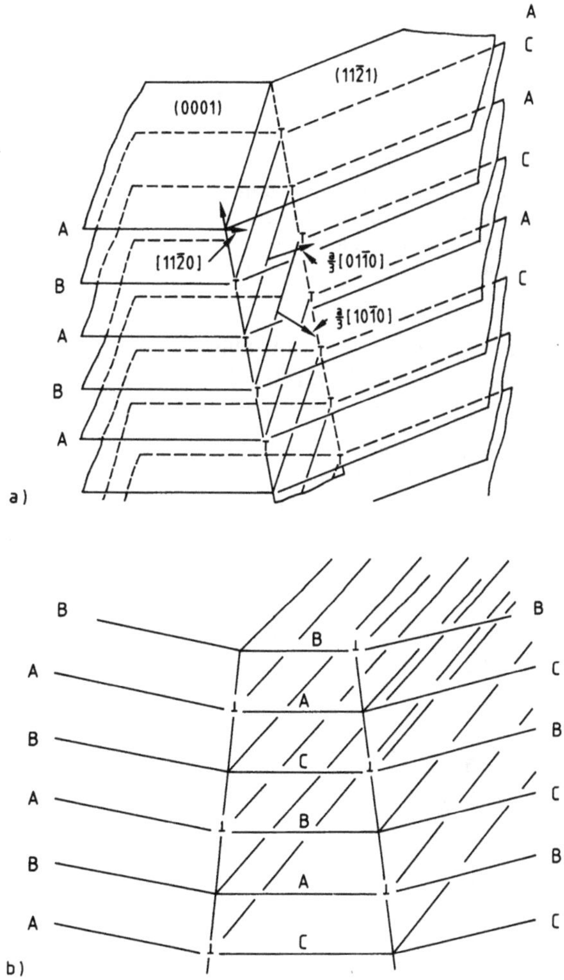

Fig. 17a, b. Formation of stacking faults with ABCABC – arrangements in graphite (Freise and Kelly[85]). **(a)** Twin-matrix boundary; **(b)** Mechanism for the formation of rhombohedral material by dissociation of a twin-matrix boundary

4 The curious finding by Matuyama[84] that rhombohedral stacking is obtained in graphite oxide formed by treating natural (hexagonal) graphite with strong oxidants is not consistent with this explanation

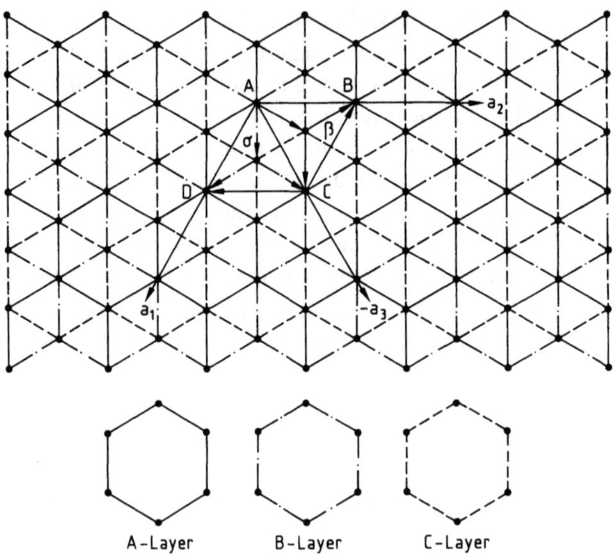

Fig. 18. (0001) basal plane of graphite for hexagonal and rhombohedral stacking sequences. The most common total dislocation with a large Burgers vector a/3 [11$\bar{2}$0] *(AC)*, lying in the basal plane, dissociates under stress into two Shockley partial dislocations with vectors a/3 [10$\bar{1}$0] (Aσ) and a/3 [01$\bar{1}$0] (σC), (after Freise and Kelly[82])

showed that severe shear introduces into graphite numerous tilt and twist boundaries as well as many twins. As further shown by Freise and Kelly[85], two twin domains with axes of tilt at 120° to one another can intersect to generate a layer with a stacking sequence ABCABC as required for rhombohedral graphite (Fig. 17). In the absence of stress, the volume fraction of rhombohedral graphite produced by this mechanism is only small. Under stress, on the other hand, a twin boundary with a (11$\bar{2}$1) habit plane can dissociate into partial dislocations of Shockley type, (a/3)[10$\bar{1}$0] and (a/3)[01$\bar{1}$0], which are shown as Aσ and σC, respectively in Fig. 18. The two partials tend to move in opposite directions under a shear stress applied in direction [1$\bar{1}$00] because forces of opposite sign are induced on their screw components. Consequently, a region is produced between the two partial Aσ and σC that comprises a low-energy stacking fault with rhombohedral stacking

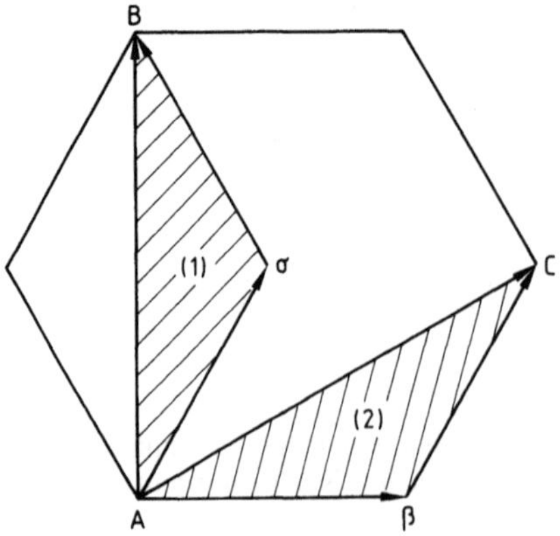

Fig. 19. Schematic representation of dislocation reactions leading to low-energy ABCABC stacking faults *(1)* and high-energy AAA stacking faults *(2)* in graphite (after Amelinckx[86])

sequence (Fig. 19, region 1). A different mechanism of dissociation of dislocations may be envisaged leading to a high-energy stacking fault with "A over A" stacking (Fig. 19, region 2) typical for the 1H polytype of graphite (Carbon VII[87]) observed during regraphitization of diamond behind a decaying shock front at high residual temperatures[5].

From the fact that high temperature treatment destroys the rhombohedral phase, Jagodzinki[88] deduced that it should be the stable form of graphite at low temperatures. However, as found by Freise and Kelly[82] and Amelinckx and Delavignette[86, 89], the stacking fault energy in the hexagonal phase is always positive at ambient temperature indicating that hexagonal graphite is the stable form.

The substantial shear introduced by shock treatment of a hexagonal graphite sample renders formation of the rhombohedral phase as precursor of diamond highly probable. However, high residual temperature in the sample generated either by inherent adiabatic shear[90], owing to irreversible collapse of the pore structure of the graphite powder employed, or intentionally by flash heating, that is necessary to provide sufficient activation energy for diamond nucleation, tends to eliminate those low-energy stacking faults with ABCABC sequence. This problem is at the very heart of shock-loading techniques to produce diamond since the existence of rhombohedral graphite was thought to be vital to overcome the kinetic barrier inhibiting the conversion of the sp^2-configuration of graphite into the sp^3-configuration of diamond[1, 68, 77]. The elusive nature of the rhombohedral graphite phase, however, made it necessary to abandon the idea of simple compression of this phase. As a result, initial collapse has been proposed of the graphite lattice to a glasslike structure with short-range order arrangement of carbon atoms in diamond configuration[91]. This glassy phase can be annealed at high temperatures within microseconds to form diamond.

4.2 Carbon and Boron Nitride Phase Diagrams

4.2.1 Carbon

The history of the development of the P,T-phase diagram of carbon reflects in a very instructive way the development of thermodynamics and phase theory during the 20th century[5].

As early as 1901, Roozeboom[94] constructed a phase diagram of carbon based on early thermodynamics that recognized the higher vapour pressure of diamond as opposed to graphite at temperatures up to the melting curves. Moreover, his idea that carbon could be melted under pressure and, when rapidly quenched, would crystallize as diamond spawned a number of unsuccessful attempts to synthesize diamond[95, 96].

Oswald's rule of successive reactions ("step rule") introduced the concept of metastability into thermodynamics that Tammann[97] included in his improved carbon phase diagram. One of the main conclusions to be drawn from his approach was that, when carbon is dissolved in iron, silver or molten silicates, quick cooling should produce

5 A review paper by Neuhaus[92] on synthesis of diamond, published just one year before the first reported successful diamond synthesis by Bundy et al.[93], is an informative example how the "can be's" and the "can't be's" of theory can be nullified by one conclusive experiment

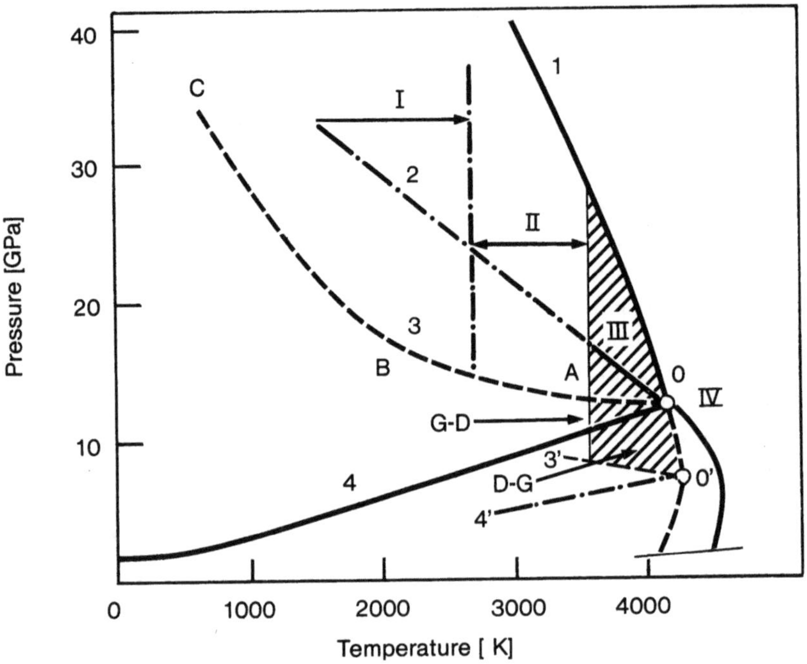

Fig. 20. Pressure-Temperature phase diagram for carbon (from Trefilov and Savvakin[116]). The position of the triple (diamond-graphite-liquid) point is shifted for ultradisperse particles from 0 to 0′. 1-melting line of diamond; 2-metastable melting line of graphite; 3-line of threshold pressure for high rates of direct graphite-diamond conversion; 3′-line of threshold pressure for ultradisperse particles; 4-graphite-diamond equilibrium line; 4′-graphite-diamond equilibrium line for ultradisperse particles; *I, II, III, IV* denote the pressure-temperature regions of direct conversion under shock compression. *D-G, G-D* denote fast diamond-to-graphite and graphite-to-diamond transformation, respectively

diamond. This assumption, however, was contradicted by the unsuccessful experiments by Moissan[98] and his followers[99, 100] [6].

Using data of heat of formation of graphite and diamond, Rossini and Jessup[103], and Leipunskii[104] determined the graphite/diamond equilibrium line that was later linearly extended to higher pressures and temperatures by Berman and Simon[105]. Experimental data provided by the first reported successful and reproducible diamond synthesis, using catalyst-solvent metals, by Bundy et al.[93] of General Electric showed that the Berman-Simon extrapolation reflects rather well the experimentally determined graphite-to-diamond transformation data[106] [7]. The position of the equilibrium line was slightly cor-

6 In an interesting twist of fate recently a "lost" paper by Burton[101] resurfaced[102] who had claimed to have produced diamond in its metastable field of stability by dissolving carbon in a lead-calcium alloy and selectively removing the calcium by steam oxidation. Repetition of Burton's experiments yielded a crystalline product that appeared to have been at least diamond-like[102]

7 There are two bodies of information indicating that formation of (metastable) diamond is possible in the region below the Berman-Simon line. The first one pertains to ion-beam deposition of thin films of diamond-like carbon (i-carbon) at low pressures in hydrocarbon gas plasma on non-diamond substrates[107–109], as well as on heated diamond seed surfaces[110–112]. The second one relies on transformation of aliphatic halogenated hydrocarbons and ring compounds such as cyclohexane and its derivatives with chair configuration at low pressures and temperatures[113, 114]

rected towards higher pressures by Kennedy and Kennedy in 1976[115]. There is evidence that the position of the Berman-Simon graphite-diamond-liquid is a function of the particle size[116]. Figure 20 shows that for ultra-disperse particles the Berman-Simon (BS) line has shifted towards lower pressures. This may indicate that some of the experiments reported to have produced diamond in its metastable region of existence below the BS-line actually have been conducted in the region between lines 4 and 4′ in Fig. 20, involving ultradisperse particles. After determination of the equilibrium graphite melting curve by Bundy[118] and Fateeva and Vereshchagin[119], of the graphite-diamond-liquid triple point[118], and of the metastable melting curve of graphite[120], a comprehensive phase diagram of carbon could be constructed[118] using also the shock-wave data obtained by DeCarli and Jamieson[1], and Alder and Christian[10] reported around this time. This phase diagram was later extended[117] and is shown in Fig. 21.

Although the position of the metastable melting curves of graphite and diamond, and the equilibrium line of graphite-to-diamond transition (BS-line) have been established

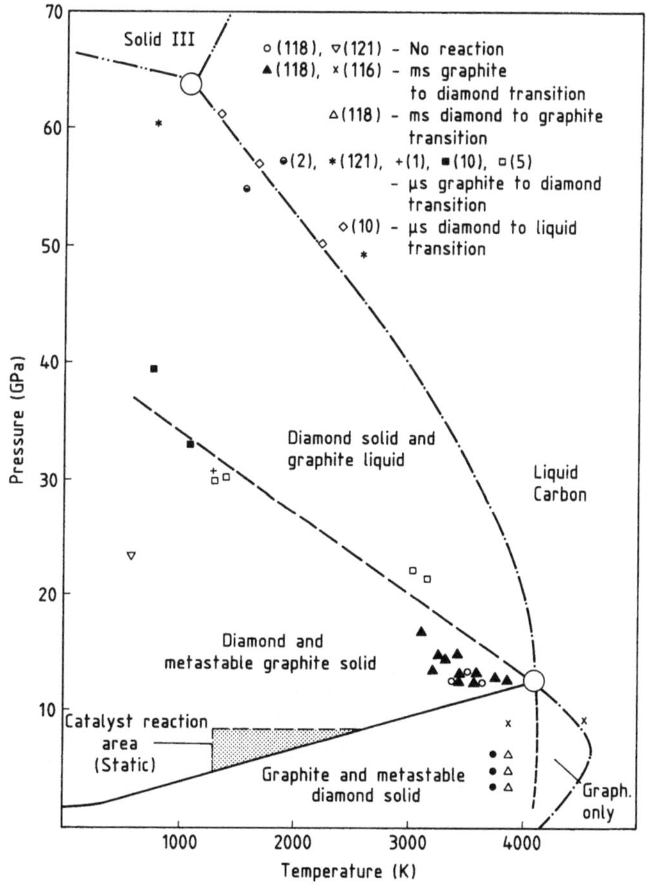

Fig. 21. Comprehensive phase transition diagram of carbon according to Bundy[117]. *Solid line:* graphite/diamond equilibrium line. *Dashed lines:* metastable melting curves of graphite and diamond, respectively. *Chain lines:* equilibrium melting curves. Experimental shock-wave data from different sources are added. Hypothetical triple point diamond/solid III/melt according to Alder and Christian[10]

with reasonable accuracy, controversial evidence exists pertaining to the high pressure/low temperature region (existence of metallic carbon) and the high temperature/low pressure region (existence of a graphite/liquid/vapor triple point, and the carbyne controversy) of the carbon phase diagram.

The field of "solid III" indicated in Fig. 21, and the position of the diamond/metallic carbon/liquid triple point have been investigated by Van Vechten[122)] based on scaling the volume changes observed for zincblende (diamond)-type to metal transitions for the Group IV and III–V compounds[8], and by Grover[127)] by applying a conservative phenomenological equation-of-state model. The phase diagrams proposed by Van Vechten and Grover are shown in Fig. 22.

Van Vechten's model[122)] involves a high-density (ϱ = 4.65 Mg/m^3) metallic solid above the diamond/metal/liquid triple point located at 3080 K and 118 GPa (Fig. 22(a)).

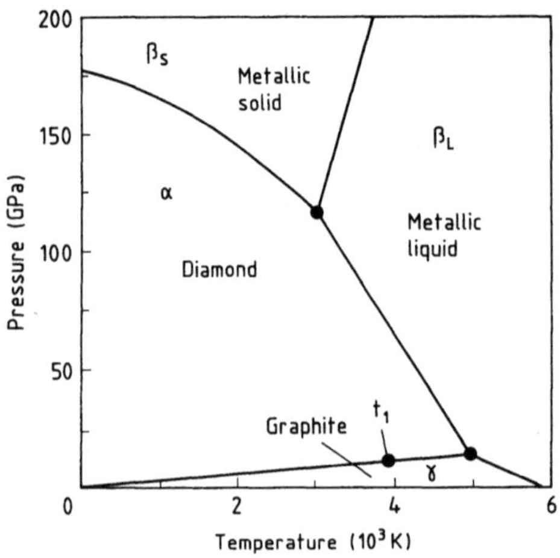

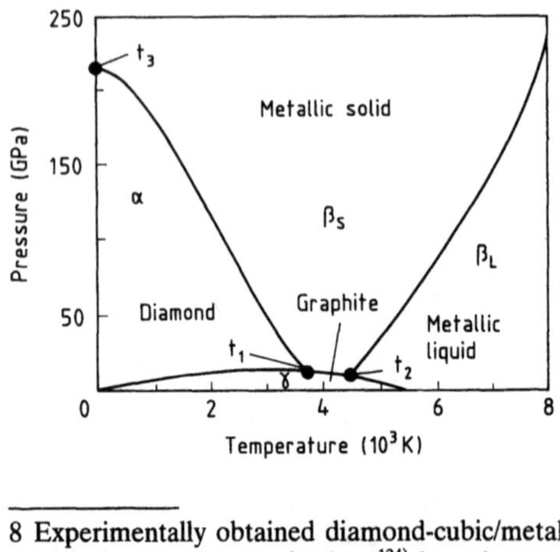

Fig. 22a, b. Phase transition diagrams of carbon involving solid metallic phases. **(a)** Diagram for a high-density metallic phase (van Vechten[122)]). t_1: experimental triple point graphite-diamond-liquid (12.5 GPa, 4000 K) according to Bundy[118)] and Bundy et al.[74)]). **(b)** Diagram for a low-density metallic phase (Grover[127)]); predicted triple points graphite-diamond-metallic solid *(t_1)*, (11.5 GPa, 3800 K) and graphite-metallic solid-metallic liquid *(t_2)*, (9.5 GPa, 4500 K). *t_3:* from shock velocity[128)] and ultrasonic data[129)] on single-crystal diamond

8 Experimentally obtained diamond-cubic/metallic/liquid triple points for Sn, Ge and Si[123)] along with the "Jamieson criterion"[124)] have been used already by Bundy[125)] to suggest the probable location of the diamond/metallic carbon/liquid triple point. Its position (approx. 2800 K and 50 GPa), however, appears to be too low compared to more recent investigations (see for example[126)]). The reason may be that linear scaling of Group IV data is insufficient to predict the behaviour of carbon in view of its exceptionally low atomic volume and high bulk modulus compared to its Group IV homologs[127)]

This means that diamond can be melted at all pressures below 120 GPa to yield a (metallic) liquid. On the other hand, Grover[127)] claims to have evidence that a metallic solid phase extends all the way down to low densities (reference density: 3.75 Mg/m^3) thus forming two closely spaced triple points (graphite/diamond/metallic solid, t_1 at 3800 K, 11.5 GPa, and graphite/metallic solid/metallic liquid, t_2 at 4500 K, 9.5 GPa), (Fig. 22(b)). As a consequence of this model, diamond could never be melted directly because the phase fields of diamond and liquid are separated by the metallic solid at pressures above approximately 10 GPa and by graphite at pressures below 10 GPa, respectively. Consequently, efforts were expended to find evidence for direct melting of diamond. Among the first experimenters to have found molten diamond in a sample of boron-doped graphite was Bundy[118)]. Likewise, Bassett et al.[130)] found evidence for melting of diamond produced from graphite as starting material. It was only recently, however, that for the first time direct melting of diamond as starting material was (serendipitously) observed at pressures greater than 12 GPa by interaction of a focused laser beam at a very high power density of 10 GW/cm^2 with a (100)-diamond anvil face[131)].

Despite claims to have observed transformation of diamond into a metallic phase at ultrahigh pressures of around 100 GPa[132, 133)], this phase remains elusive. Shock compression experiments on diamond at approximately 200 GPa yielded a maximum post-shock density of 3.95 Mg/cm^3, intermediate between the values of the metallic solid postulated by Grover and Van Vechten and thus provided some support for the still hypothetical shock-induced transition of diamond to a metallic phase[134)]. Discontinuities found in the shock-velocity vs. particle-velocity plots at shock pressures to 120 GPa were taken as evidence of a solid-to-solid phase transition from diamond to a metallic state of carbon[135)].

As mentioned above, there are unsolved problems surrounding the nature of the high temperature/low pressure region of the carbon diagram. It is not known yet whether there exists a true graphite/melt/vapour triple point. Derjagin et al.[136)] and, above all, Kasatochkin et al.[137)] and Whittaker[138)] proposed the existence of an interesting class of allotropic modifications of carbon with linear chains of doubly or triply bonded atoms (carbynes) as transformation product of graphite at temperatures above 2500 K and low pressures. According to Whittaker[138, 139)] the compound that actually melts is carbon VI[140)], and the triple point solid/liquid/vapour is at 3800 K and 2×10^4 Pa[141)], (Fig. 23). There is, however, controversy whether these linear polytypes of carbon actually exist[9].

On the other hand, Basset[142)], Gokcen et al.[143)] and other workers observed actual melting of graphite at approximately 4150 K in atmospheres of inert gases above a pressure of 11 MPa ("110-atm faction" of Bundy[117)]). It is difficult to dismiss this evidence. It is likewise precarious to push aside the mounting evidence for the existence of chain-like carbon allotrops. To complicate matters even more, a new structural polytype of polymeric carbon, layer-chain carbons with 2/3 of the atoms characterized by sp^3-hybridization (diamond-type) and 1/3 by sp-hybridization (carbyne-type), was identified quite recently[153)]. The structure of layer-chain carbons is intermediate between graphite and diamond in that they differ from the former by having covalent bonds between the layers, and from the latter by having two long C–C bonds linking the layers. The general notation of the structures is mLCp, where m denotes the number of layers and p denotes

9 An in-depth discussion of carbynes is beyond the scope of this review. The reader is referred to[144–148, 150–152)]. See also: Notes Added in Proof

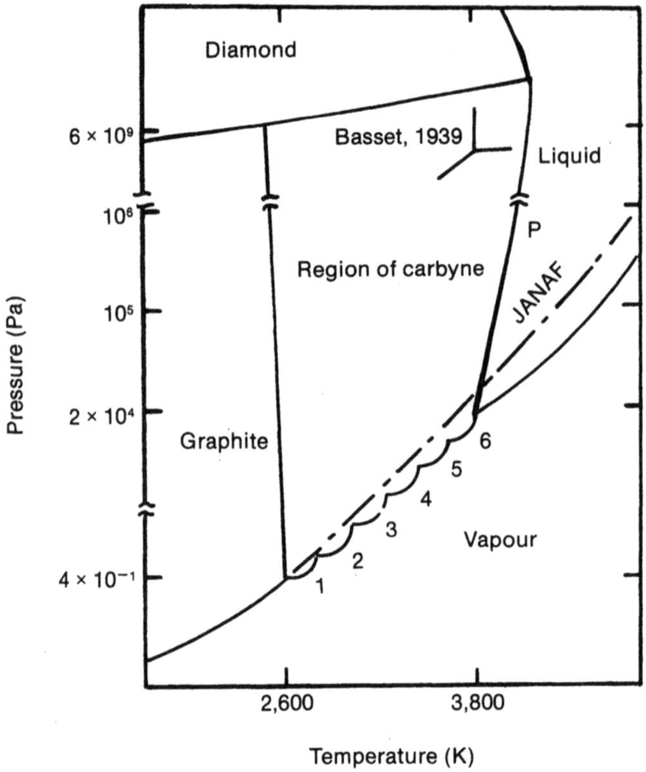

Fig. 23. The high temperature/low pressure region of carbon as proposed by Derjagin et al.[136)] and Whittaker[138)], and the position of the solid/liquid/vapour triple point according to Basset[142)] and Gokcen et al.[143)]. The numbers *1–6* refer to different proposed carbyne forms

the length of the carbon chains[10]. As suggested by the authors[153)], interplanar spacings additional to graphite observed earlier by Franklin[154)] during thermal conversion of poorely- and non-graphitizing carbon could be explained in terms of formation of layer-chain carbons at temperatures exceeding 2300 K.

Similar hybride structures seem to exist in glassy carbon. According to Noda and Inagaki[155)] glassy carbon contains planar carbon-carbon bonds of graphite type, and tetrahedral bonds of diamond type, (see also[156)]).

In conclusion, the carbon phase diagram is reasonably well known at temperatures up to 3000 K and pressures up to 45 GPa. Poor pressure control in ultra-high pressure experiment to pinpoint the location of the triple point diamond/metallic phase/liquid, and to prove the existence of the metallic carbon phase have hampered research in this area. On the other hand, experimental difficulties have been encountered in terms of accurate temperature measurements at the temperatures around 4000 K required to search for the exact triple point graphite/liquid/vapour. The unsettled question of the existence of carbynes, and the proposed existence of the newly discovered layer-chain carbons in this PT-

10 Thus, graphite is mLC0 (2LC0 = hexagonal graphite, 3LC0 = rhombohedral graphite; zero indicating the absence of interlayer links), diamonds are mLC1 (2CL1 = lonsdaleite, 3LC1 = cubic diamond). The class mLC2 is called carbyno-diamonds, with 2LC2 = two-layer (lonsdaleite) carbyno-diamond and 3LC2 = three-layer (cubic diamond) carbyno-diamond

region add to the confusion. How many surprises the carbon system may still hold is vividly illustrated by the recent discovery of spherical shell molecules with only sp^2 hybridized carbon atoms, such as "footballene (soccerballene)", C_{60} and "Archimedene", C_{120}[149].

4.2.2 Boron Nitride

Boron nitride with zincblende structure (z-BN, "Borazon") was first obtained by Wentorf[157], from hexagonal-sheet type boron nitride (g-BN) treated under static pressure conditions using catalysts such as alkali and alkaline-earth metals, tin, antimony, or lead. Direct transformation without a catalyst of g-BN to w-BN at temperatures below 2300 K, and to z-BN at temperatures between 2500 and 4000 K at minimum static pressures of 13 GPa was reported by Bundy and Wentorf[77]. Later, Adadurov et al.[6], DeCarli[158] and Coleburn and Forbes[9] synthesized high-density modifications of boron nitride by shock compression techniques. Discontinuities found in the shock-velocity vs. particle-velocity plot ("shock Hugoniot") at $U_s = 5.037$ km/s, $U_p = 1.137$ km/s and $U_s = 5.30$ km/s, $U_p = 1.93$ km/s were interpreted as onset and end of the transition of g-BN into a dense form of boron nitride[9]. Figure 24 shows the shock-Hugoniot and the pressure-volume curve (inset) derived from the shock data. The latter curve shows a cusp at 12.2 GPa which is the minimum transition pressure. This value agrees well with shock compression-derived transition pressure data of 12 GPa by Alt'shuler et al.[159], Adadurov et al.[6] and Dulin et al.[16], and static compression-derived data of 11.5 GPa at 2000 K by Bundy and Wentorf[77] and Corrigan and Bundy[78].

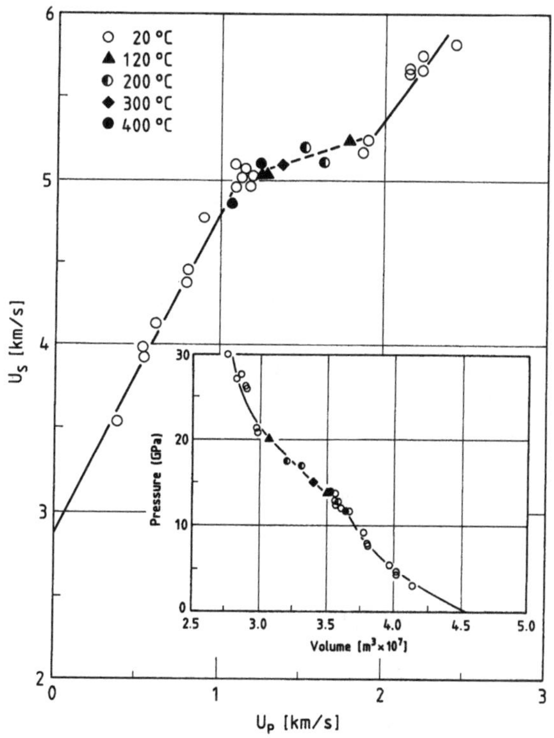

Fig. 24. Shock Hugoniot of boron nitride and derived pressure-volume relation (inset). U_s = shock velocity, U_p = particle velocity (after Coleburn and Forbes[9])

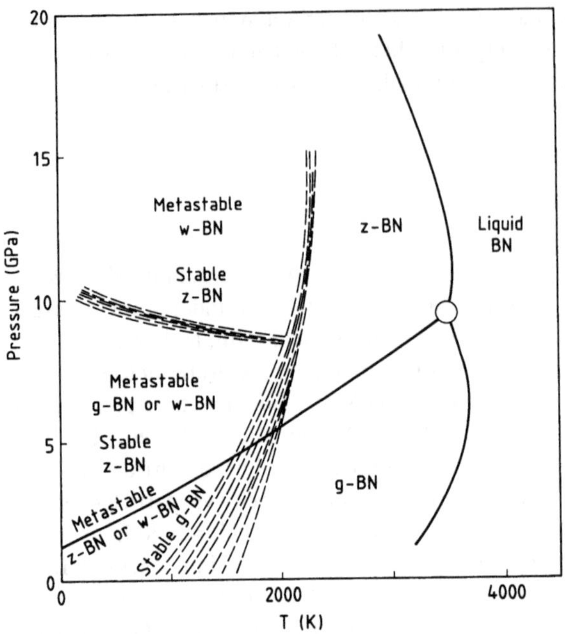

Fig. 25. P,T phase and reaction diagram for boron nitride (after Corrigan and Bundy[78])

The nature of the high-density phase obtained by shock loading, however, appears to be controversial. The results by Adadurov et al.[6], Soma et al.[62] and Sawaoka et al.[160] indicate formation of w-BN above approximately 23 GPa, whereas Coleburn and Forbes[9] and Dulin et al.[16] found z-BN at comparable pressures. This discrepancy, as suggested by Gust and Young[161] could be explained by the presence of catalyzing impurities in the starting material used by the latter authors[11]. It is known from experiments by Bundy and Wentorf[77], Nikol'skaya et al.[71] and Susa et al.[162] that addition of catalysts always leads to z-BN. Moreover, as pointed out by Corrigan and Bundy[78], w-BN is not a thermodynamically stable phase in the boron nitride system. Thus, metastable w-BN can form only from (metastable) g-BN at low temperatures (see Fig. 25) by the athermal martensitic transformation mechanism outlined in 4.1.3.2. Formation of z-BN from g-BN or w-BN requires either high temperatures to provide the very high energy necessary to disrupt the lattice in order to rearrange it in a z-BN configuration[12], or the use of catalysts to lower the activation energy for the disruption/rearrangement mechanism.

The stability of w-BN under high pressure has been investigated by Tani et al.[163] and Corrigan and Bundy[78]. Figure 26 shows the results obtained from post-compression X-ray diffraction analyses.

Figure 25 shows the complete low-pressure phase diagram of boron nitride as determined by Corrigan and Bundy[78] in 1975. As can be seen by comparison with the carbon phase diagram (Fig. 21) the behaviour of the boron nitride system appears to be similar (see also[164]).

---

11 Moreover, the lattice constant of z-BN (0.361(6) nm) is almost identical to that of copper (0.360(7) nm). Thus formation of z-BN reported in experiments involving copper sample holders should be viewed with skepticism[160]

12 The activation energy for the conversion of g-BN → z-BN was estimated to be about 600–1050 kJ/mol at 6.5 GPa, and for w-BN → z-BN to be about 850 kJ/mol[78]

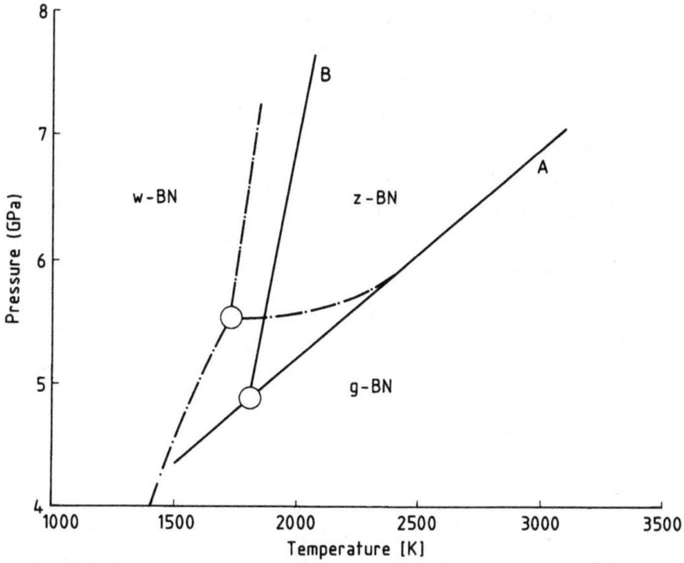

Fig. 26. Phase relation of boron nitride polymorphs. *Solid lines A* and *B* according to Corrigan and Bundy[78)], *chain lines* according to Tani et al.[163)]. The triple point *w-BN/g-BN/z-BN* is at approximately 1780 K/4.85 GPa and 1725 K/5.5 GPa, respectively

4.3 Interaction of Shock Waves with Graphite and g-BN

The severe conditions in terms of pressure and temperature imposed on a crystal lattice by shock compression might suggest that the ordered lattice undergoes chaotic destruction at the atomic level ("Götterdämmerung" view, according to Johnson and Mitchell[165)]). However, many successful phase transformation experiments involving shock waves have shown that they proceed in an orderly fashion and induce crystal structure transformations within 10^{-7}–10^{-9} s. This has been deduced solely from the results of shock-quenching experiments until Johnson and Mitchell[165)] in 1972 for the first time obtained an X-ray diffraction pattern of shock-transformed pyrolytic boron nitride during compression.

How the transformation process proceeds in detail is still a matter of conjecture, and several hypothetical mechanisms have been advanced ever since the first successful diamond synthesis under dynamic conditions, without catalysts or seeds, was achieved by DeCarli and Jamieson[1)]. The structural aspects of shock-induced transformation of graphite and boron nitride have been described in the comprehensive book by Kurdyumov and Pilyankevich[164)].

4.3.1 Lattice Compression

According to DeCarli and Jamieson[1)], a simple lattice compression in c-axis direction of rhombohedral graphite (Fig. 14(b)) would be required to produce diamond by nondiffusional processes. Because the hexagonal carbon sheets in graphite are planar, but puckered in diamond, out-of-plane displacement of carbon atoms is required to yield the chair-form arrangement in diamond (Fig. 14(d)). An always present shear stress component in shock waves[43, 166] could easily account for the small displacement vector of only 0.25 Å required for puckering. This mechanism had initially been suggested by Lonsdale et al.[17], and was also considered by Lipschutz[167] for the solid-state diffusionless conversion of graphite to diamond found in meteorites.

Calculations of compressibilities of graphite and diamond using molecular force constants for carbon-carbon stretching and out-of-plane displacement, performed by Riter[72] showed, however, that out-of-plane buckling or puckering is unlikely below pressure of 35 GPa. It appears that graphite bond compression is the leading mechanism to absorb interplane compressive energy below this threshold pressure. Addition of a catalyst, however, might "dump" in-plane compressive energy into puckering modes. This mechanism was supported by Coulson and Haigh[168] who found that a ring system under certain circumstances will tolerate a considerable amount of compressive stress, and still remain planar. Upon exceeding a threshold value the excess energy stored in bond compression will be spontaneously released: the molecule will be distorted out of its plane.

Contrasted with the nondiffusional process of lattice distorsion should be a mechanism suggested by Chomenko et al.[169] that relies on self-diffusion of carbon atoms in a graphite lattice[170]. Figure 27 shows that not only in a-direction (Fig. 14(d)), but also in c-direction a chair-like (puckered) diamond hexagon can be formed. Movement of carbon atoms into interstitial positions occurs readily at high pressures when the graphite lattice becomes unstable. It has been shown[169, 171] that the most likely position of an interstitial atom is a site which could be projected onto the center of a hexagon, and which is equidistant from all atoms of this hexagon, and the next nearest one lying in a plane above (Fig. 27). As emphasized by Mogilevskii[172], "dissolution" of lattice atoms of planes perpendicular to the direction of shock wave propagation takes place under sufficiently high pressures. Those liberated atoms can move freely through the lattice forming clusters of interstitials in the planes close to the normal to the shock wave front. Thus, relaxation of elastic stresses imposed by the transition of the shock front through the lattice is not achieved by dislocation glide but rather by a non-conservative movement of boundary dislocations. The process is a rapid mechanism with small activation energies of the order of 0.1 eV[173].

If the simple intuitive approach of compression of rhombohedral graphite[1] were correct, deliberate addition of rhombohedral graphite to the initially hexagonal graphite

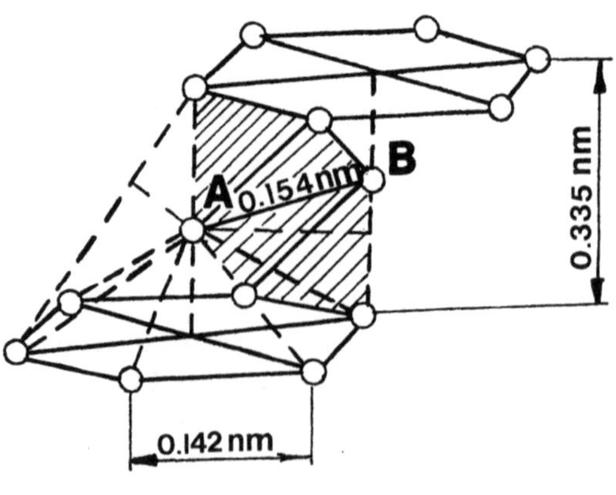

Fig. 27. Formation of the characteristic bond length *(AB)* of a "puckered" diamond hexagon by injection of interstitial carbon atoms *A* and *B* between planar graphite layers (after Chomenko et al.[169])

should substantially increase the yield of diamond. This, however, was never observed. Following Delavignette and Amelinckx[89] (Fig. 28), only a few lamellae of rhombohedral abcac-stacking, inserted into the bulk crystal with normal hexagonal abab-stacking suffice to relax, without breaking C–C bonds, the elastic stresses generated, for example, by mechanical grinding[79, 81]. Thus, the relative proportion of graphite with rhombohedral stacking order always appears to be small. Moreover, increasing temperature necessary to provide the activation energy for the reassembly of the diamond lattice destroys the rhombohedral phase quickly[88].

4.3.2 *Glassy Transitional Phase and "Hot Spots"*

The failure to explain the mechanism of shock conversion of graphite into diamond by a nondiffusional compression mechanism led DeCarli[91] to conjecture an instantaneous collapse of the graphite lattice to a glasslike structure with short-range order arrangement of carbon atoms in diamond configuration. Sufficiently high shock temperatures then can lead to annealing and recrystallization, respectively of the glassy phase within microseconds. This process may continue in the wake of the rarefaction wave as long as high enough post-shock temperatures are available. The mechanism implies that diamond yield should be a function of the duration of the pressure pulse which increases the amount of glassy phase. Experiments performed with pulse interaction times between 0.5 and 10 μs indicated, however, that the yield was essentially independent of the duration of the pressure pulse[91].

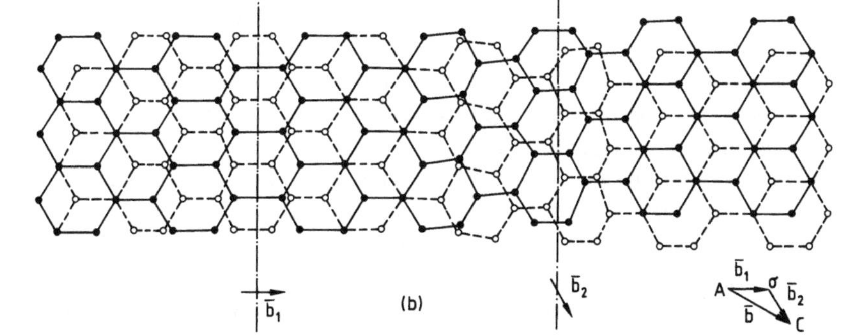

Fig. 28. (**a**) Formation of a lamella of rhombohedral graphite *(abcac, center)* between areas with hexagonal graphite symmetry *(abab, left* and *right)*. (**b**) Projection onto (0001). The dislocation with Burgers vector $\mathfrak{b}$ = (AC) splits into two Shockley partials with a Burgers vector $\mathfrak{b}_1$ of screw-type (Aσ) and a mixed $\mathfrak{b}_2$ dislocation (σC). The area with deformed hexagons of rhombohedral stacking order is enclosed between the partials thus comprising a stacking fault (Delavignette and Amelinckx[89])

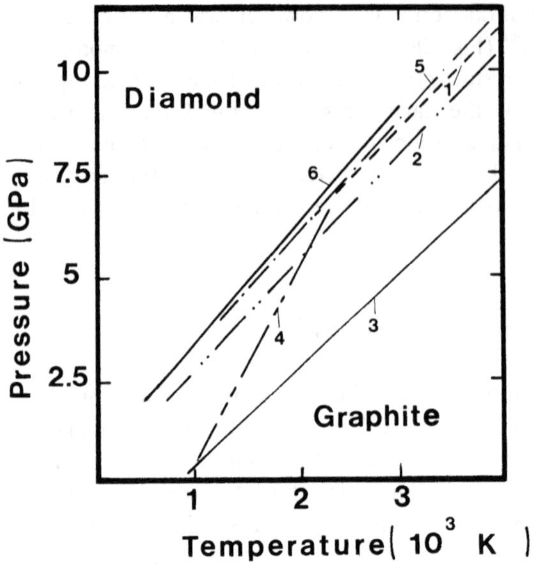

Fig. 29. Equilibrium lines of diamond with various carbon materials (from Vereshschagin et al.[182]). *1* – graphite (monocrystal); *2* – pyrolytic carbon; *3* – glassy carbon; *4* – coke; *5,6* – different grades of graphite

These results required readjustments of the theory, and in turn DeCarli proposed the existence of "hot spots" within the bulk graphite to be transformed into diamond[174]. Hugoniot equation of state (EOS) data of graphite[175] showed that porous graphite transforms more readily into diamond than nonporous material[176], (see Fig. 29). The reason for this is not immediately obvious. It may be that the very complex response (see Fig. 5) of a porous material to shock compression causes a deviation from the Hugoniot EOS measured in single crystalline material, that leads to nonequilibrium thermodynamic states. This thermodynamic condition of instability can be based upon a principle of positive-entropy production for the irreversible collapse of the initial compaction surface[177]. This view is closely associated with the assumption of saturation of the solid with dislocations at the onset of adiabatic shearing or heterogeneous melting. DeCarli's[174] "hot spots" are created in pressure nodes and their existence may be either attributed to jetting and adiabatic shear[90] by collision of micro-jets and cavity walls of collapsing pores[178] or to reverberation of the shock front[179] and plastic deformation work during pore collapse[180, 181]. The peak temperatures are likely to exceed 3500 K, and can be estimated by the simple relation $T_s \propto U_p^2$, where U_p equals the particle velocity behind the shock front[179].

Figure 29 shows that the position of the transition line (graphitic) carbon to diamond (Berman-Simon line) depends on the structural order and the density of the starting material (see also[178]). Porous coke requires much lower conversion pressures at low temperatures. This may be consistent with the "hot spot" approach in porous materials. The glassy carbon may already contain tetrahedrally-bonded diamond-like or carbyne-like carbon atom configurations[155], which would act as nuclei for diamond formation upon shock loading.

The adiabatic shear produced by collision of micro-jets with cavity walls of the collapsing pores gives rise to shock-induced high temperatures. Figure 30 shows that these temperatures increase with shock pressure. Moreover, lower densities (higher porosity) of the graphite powder to be compressed generally results in higher temperatures.

The mechanism of "hot spot" formation may further provide an interesting situation as far as the dissipation of thermal energy is concerned. At the very high shock pressures encountered in the hot spots, thermodynamic self-cooling can take place resulting from adiabatic expansion of matter in the rarefaction wavefront. Cooling then is effected through fast loss of internal energy by performing work against the shock pressure applied from outside, and not by slow conservative heat transfer by either conduction or radiation. This means that some diamonds formed in "hot spots" are quenched rapidly below the Berman-Simon line thus avoiding regraphitization or mechanical stresses introduced by high residual temperatures or steep temperature gradients. The mechanism of adiabatic cooling may well be responsible for the preservation of natural diamonds formed in a kimberlitic magma[13] by cavitational processes involving carbonaceous gases. According to Bernoulli's principle, these gases may be highly pressurized, while approaching a constriction along the path of the explosively extruded magma within a diatreme[184].

In summary, pressure and temperature conditions in "hot spots" deviate strongly from the average P-T conditions in the bulk sample. Then, diamond nucleation can occur in "hot spots" even if the remainder of the sample did not reach the transformation conditions, i.e., if it remained below the Berman-Simon line.

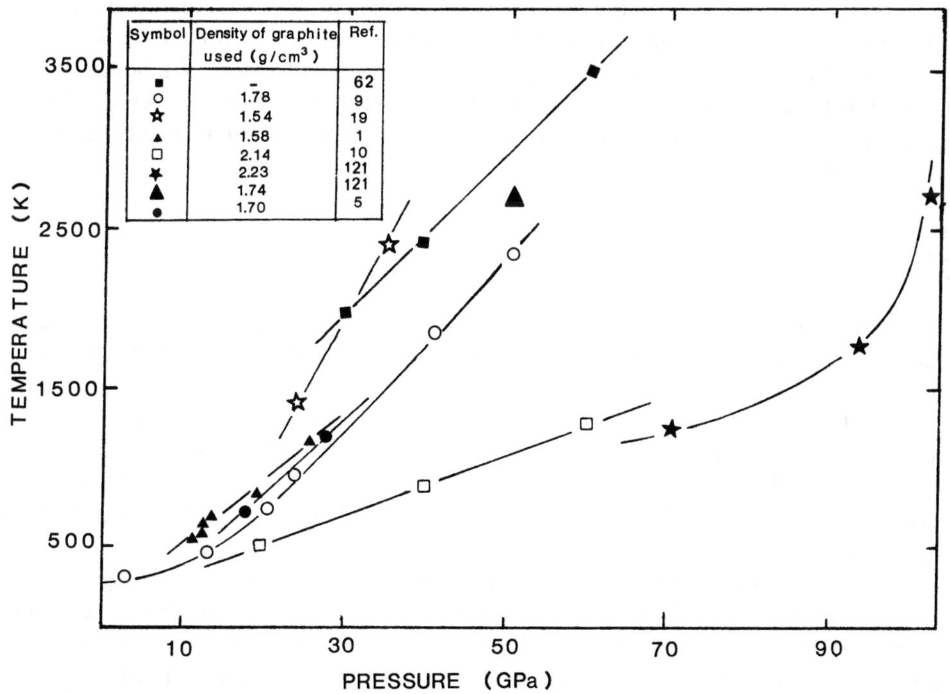

Fig. 30. Shock-induced temperatures versus shock strength in graphites with different initial densities (from[183])

13 There are, however, compelling reasons to assume that diamond formation did not take place in a kimberlitic or lamproitic magma[61]

4.3.3 Formation of Liquid Phase

The mechanisms discussed so far treat the conversion of graphite to diamond in the wake of a shock wave as a real solid-to-solid transformation.

Morris[19)], however, pointed out that "hot spots" are able to convert only a small fraction of graphite to diamond at a given shock intensity. Moreover, to explain the high growth rates observed in shock experiments, he suggested a very high density of different lattice defects introduced by the passing shock pulse that leads to a dramatic increase in the rate of diffusion of carbon atoms. This is well in accord with the findings of other researchers[18, 185)], and is also reconcilable with the analysis performed by Horie[177)] of melting of solids under shock compression[14]. He explained melting phenomena by combining dislocation theory of melting with the thermodynamics of formation of dislocations. Then, adiabatic shearing takes place in a solid under shock compression by catastrophical dislocation multiplication. If one considers melting as large shearing displacement[187)] then a melt is essentially a solid saturated with dislocations.

Consequently, liquid or quasiliquid layers between graphite and shock-produced diamond can enhance not only growth rates of diamond but may also act as promoters for nucleation of diamond by decreasing the Gibbs free energy of formation of diamond nuclei. This liquid layer then fulfills the same purpose as the catalyst in static pressure diamond synthesis. However, whether a true liquid phase appears during shock conversion of graphite is still a matter of speculation. The pronounced spherical shape of diamond or diamond-like inclusions in shocked graphite found by Kleiman et al.[5)] and also observed by Balan et al.[188)] strongly suggested the involvement of a liquid phase formed by the combined action of a flash-heating pulse and the adiabatic heating of the graphite (see above). This solid-vapour-liquid-solid (SVLS) theory of diamond formation is dealt with in Sect. 4.3.5.

4.3.4 Dislocations and "Weak Spots"

Lattice defects such as dislocation lines, vacancy clusters, stacking faults, interstitial atoms and the like represent "weak spots" which render a crystallographic pattern different from that of the remainder of the structure. Distortional energy stored in weak spots can be released during lattice relaxation in the wake of a shock wave, and may give rise to a substantial energetic contribution to the total Gibbs free energy required for formation of diamond nuclei. As a consequence, nucleation of diamond could occur at weak spots in a graphite lattice even if the total system is located in P,T – plane below the graphite-diamond equilibrium line[182)].

Pujols and Boisard[18)] conjectured that an extremely large number of stacking faults is formed due to plastic deformation behind a shock front suggesting a martensitic process for graphite-diamond transformation. These stacking faults form lamellae of rhombohedral stacking order, and this formation mechanism involves dissociation of basal dislocation with large Burgers vectors into an array of Shockley partial dislocations[82, 89, 189)]; (Fig. 28).

14 Sublimation of NaCl on impact of a shock wave has been experimentally observed as early as 1956[186)] (see also[21)])

In this way, "weak spots" are created that have the character of defects with a cubic closed-packed stacking order, and that move ahead of the shock front. It was shown by Solov'ev[190] that under the action of a shock wave stacking faults can spread and propagate. The propagation can involve climb and step formation if, for example, an advancing stacking fault reacts with a "forest" of screw dislocations (Fig. 31(a)). Steps can multiply by coalescence and branching at a partial-dislocation source (Fig. 31(b)) which causes the stacking faults to occupy different atomic planes. By this mechanism a larger volume is created with c.c.p. atomic arrangement.

Movement of Shockley partials in basal planes of graphite occurs readily without the need of high pressures. The temperature excerted by the beam of an electron microscope on the sample appears to be sufficient to induce dislocation movement. Figure 32 shows the fusion of two ribbons of partial dislocations marked with an arrow in panel A. The micrograph of panel B was taken 3 s after that of panel A. Panel C shows the explanation of this fusion reaction according to Delavignette and Amelinckx[89]. On meeting, the two partials σB and Bσ annihilate each other, and a single ribbon Aσ + σC (Fig. 19) results. The interior of the curved triangle in Fig. 32(c) contains a stacking fault of the same type as the generating ribbons, i.e., of rhombohedral stacking order[183].

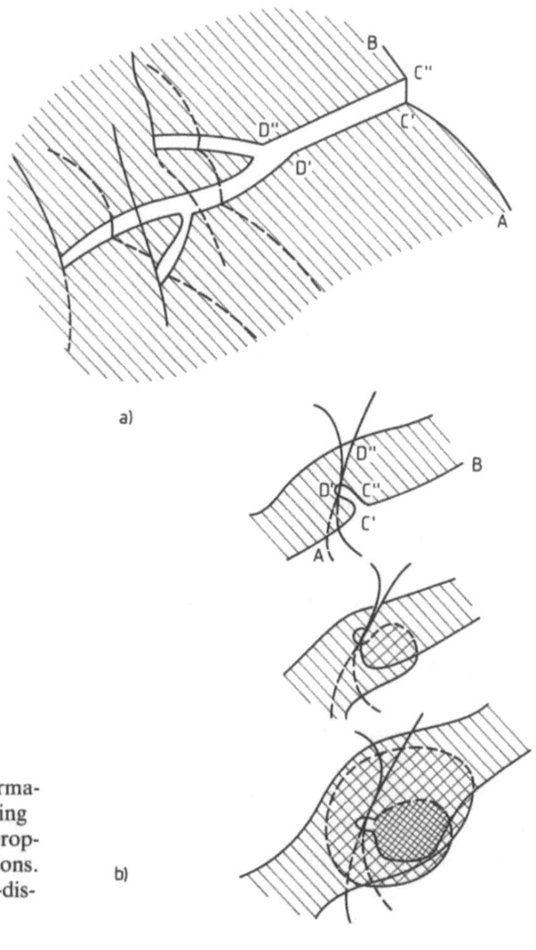

Fig. 31a, b. Dislocation climb and step formation (according to Solov'ev[190]). **(a)** *A* stacking fault between the partials *C'D'* and *C''D''* propagates through a "forest" of screw dislocations. **(b)** Formation of multiple steps at a partial-dislocation source

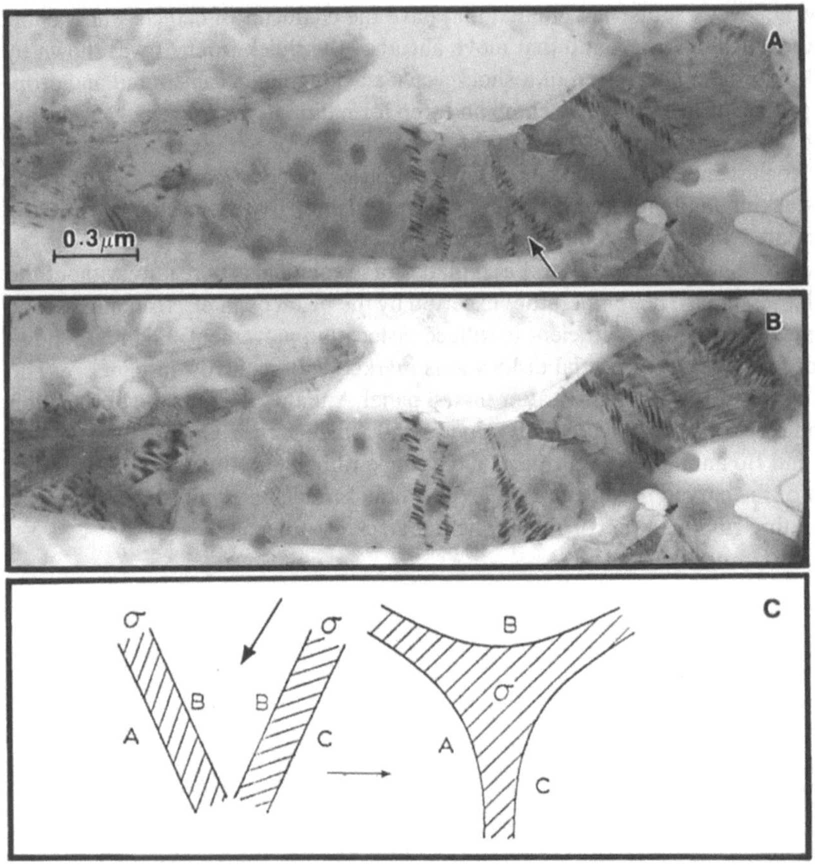

Fig. 32A–C. Two consecutive electron micrographs (**A** and **B**) of a graphite particle, showing fusion of dislocation ribbons, and the schematic representation of the observed phenomena (**C**). (After Delavignette and Amelinckx[89])

Even if a quasi-critical diamond nucleus has been formed, it has to be ensured that it grows and also remains stable upon release of pressure and temperature. However, a single shock wave impulse might be too short lived to cause a permanent rearrangement of the graphite lattice[191]. It may cause only a vigorous oscillation of carbon atoms with sp^2 configuration towards the sp^3 configuration of diamond. After traverse of the shock front, the "prediamond" cell returns quickly to its original state. Adjacent graphite cells do not "know" the shock history of their neighbours, and no cooperative phenomena can be observed.

Clearly, one has to find a way to prolong the interaction time of lattice and shock wave. This can be done either by retention of the shock compressed material in rigid ampoules[192] or by addition to graphite powder of fine metal particles that provide shock impedance mismatch between the host lattice and the weak spots within the lattice. The advancing shock wave pulse will be trapped in areas of weak spots, reverberate and increase the interaction time. Neighbouring cells will now be cooperatively involved, and their wave functions can overlap long enough to create coupled bonding states and, in turn, stable nuclei. In addition, rarefaction waves generated in the wake of the reverberating shock front prolong the time graphite particles are subjected to high pressure, and protect the diamond nuclei from being regraphitized by quenching.

A beneficial effect of the metal addition is the temperature reduction in the sample which prevents reconversion of diamond or w-BN[2, 193, 194]. The "weak spot" approach and the "hot spot" concept as described in 4.3.2 are mutually complementary. While the hot spot concept tries to explain diamond formation by shifting the P-T conditions within a hot spot into the field of stability of diamond, the weak spot concept describes diamond formation by creation of stacking faults with rhombohedral stacking order whose distortional energy released behind the shock front provides the activation energy for diamond nucleation.

4.3.5 SVLS Model

The solid-vapour-liquid-solid (SVLS) model[5, 195] of formation of diamond in shock-compressed graphite tries to reconcile discrepancies found between the rate of nucleation predicted by the hot spot approach, and the observed yield of diamond[19, 174]. The pronounced spherical shape of diamond and regraphitized particles found in shock experiments performed by Kleiman et al.[5] strongly suggested the presence of a liquid phase formed by the combined action of the flash-heating pulse applied and the adiabatic heating by collapse of the pore walls of the graphite particles. Figures 33 and 34 show typical examples of round inclusions in graphite. Sometimes the round particles tend to agglomerate[196]. Figure 35 shows an aggregate of carbon particles with decidedly different structural states. The existence of grains with diamond-like[15] (Fig. 35(a)) and cubic diamond (Fig. 35(e)) structures in close proximity to turbostratic graphite (Fig. 35(b)) and amorphous carbon ("carbon black") (Fig. 35(d)) indicates very steep pressure gradients (pressure "pockets") over regions of hundreds of nm^2.

The Kleiman-Heimann-Salansky-model of diamond formation by a SVLS sequence proposes four consecutive steps: (i) formation of liquid droplets of carbon by condensation of gaseous carbon in pore spaces[198], (ii) nucleation of diamond by freezing the droplets at high pressures and high temperatures, (iii) growth of diamond, and (iv)

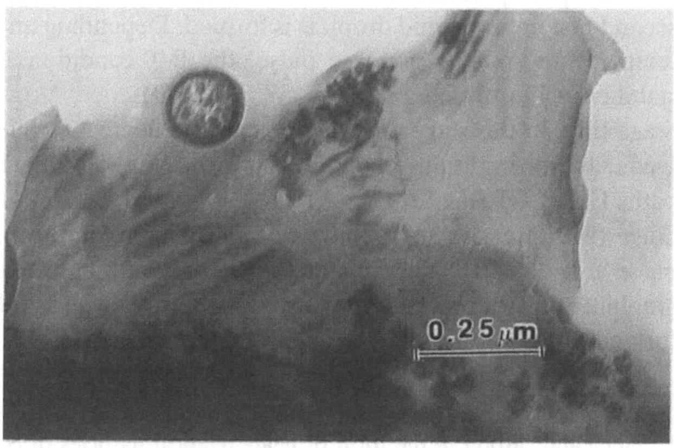

Fig. 33. Round diamond particle and "ghosts"[197] formed by interaction of a shock wave with a flash-heated graphite/copper sample

15 This diamond-like carbon polymorph has been tentatively labeled "p-diamond"[5]

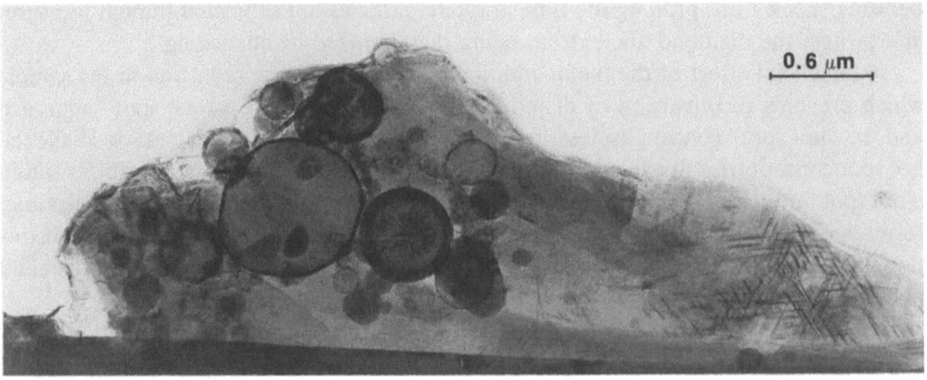

Fig. 34. Round "regraphitized" diamond particles consisting of various carbon polymorphs

regraphitization of diamond and recrystallization of carbon blacks[16]. The panels A to D of Fig. 36 refer to the four stages mentioned above.

Stage (i): Some graphite material evaporates on application of a high-temperature pulse at ambient pressure, and forms a small pocket of gaseous carbon within the pore spaces[202)], Fig. 36 (A). At sufficiently high vapour pressure, liquid droplets can form at sites or preferential nucleation ("weak" or "hot" spots). Samples with higher porosity are expected to reach higher temperatures because of their higher resistance to current flow[17]. The larger number of voids favour the formation of gas-filled micropockets. Figure 37 shows the superposition of three so-called dynamic diagrams[203)] on a carbon phase diagram. The P-T conditions of the first stage of liquid nucleation will lie on the line OA in the high temperature region. Since a negative surface free energy of graphite has been predicted at temperatures above 2500 K[204, 205)], individual crystal layers and layer aggregates would be expected to detach spontaneously above this temperature. Ejection of small particles from the surface of graphite actually has been observed at temperatures of 2950 K[198, 206)].

Stage (ii): A pressure pulse is applied to the preheated sample which forces the gas-filled pockets of carbon to collapse (Fig. 36(B)). The volume of gas in the pockets decreases and its vapour pressure increases drastically again. On reaching saturation, a number of second-generation liquid droplets is formed. Depending on the time interval between the temperature and the pressure pulse, the P-T conditions in the sample can reach the stability region of diamond (Fig. 37, path FG).

Stage (iii): At this stage the pressure starts to decrease while the temperature is still high. Under these conditions diamond can grow (Fig. 36(C)). The dynamic diagram follows paths CE or GI (Fig. 37).

Stage (iv): Intensive regraphitization of diamond and formation of carbon blacks takes place (Fig. 36(D)). Very characteristic spherulitic structures with rims consisting of graphite and different carbyne forms (Fig. 38), as well as fully graphitized spherulites are formed (Fig. 34). This stage is represented by the paths KO or MO in Fig. 37.

16 The term "carbon black" refers to a group of industrially produced materials composed of elemental carbon in the form of near-spherical particles of colloidal size[199, 200)]. They are produced by partial combustion or thermal decomposition of hydrocarbons[201)]

17 See Chap. 5 for details on the experimental set-up

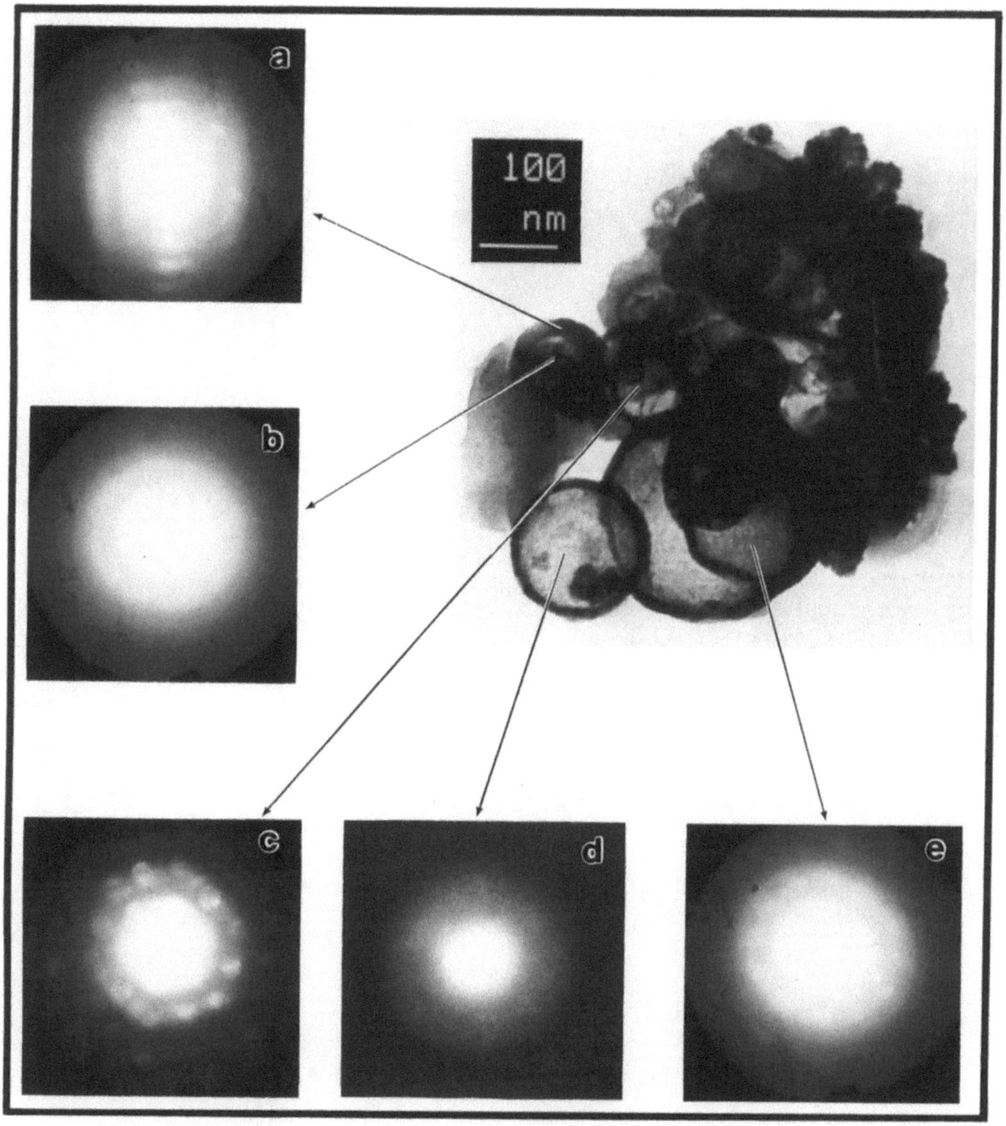

Fig. 35a–e. Scanning transmission electron micrographs of an aggregate of carbon polymorphs formed in a flash-heating implosion experiment[196]. **(a)** to **(e)** show selected area diffraction (SAD) patterns of the different regions indicated by arrows. For explanation see text

It has been shown exerimentally[5], that the degree of evaporation and the number of droplets formed, respectively, at stage (i) are proportional to the energy of the preheating pulse. Moreover, the higher the pressure applied during stage (ii) the faster carbon vapour pressure will reach saturation, and more droplets are formed. The ratio of droplets formed during stage (ii) to those formed during stage (i) is approximately 3[183].

Finally, the number of diamond nuclei formed is larger at small time intervals between the temperature and pressure pulses than at long time intervals. In the latter

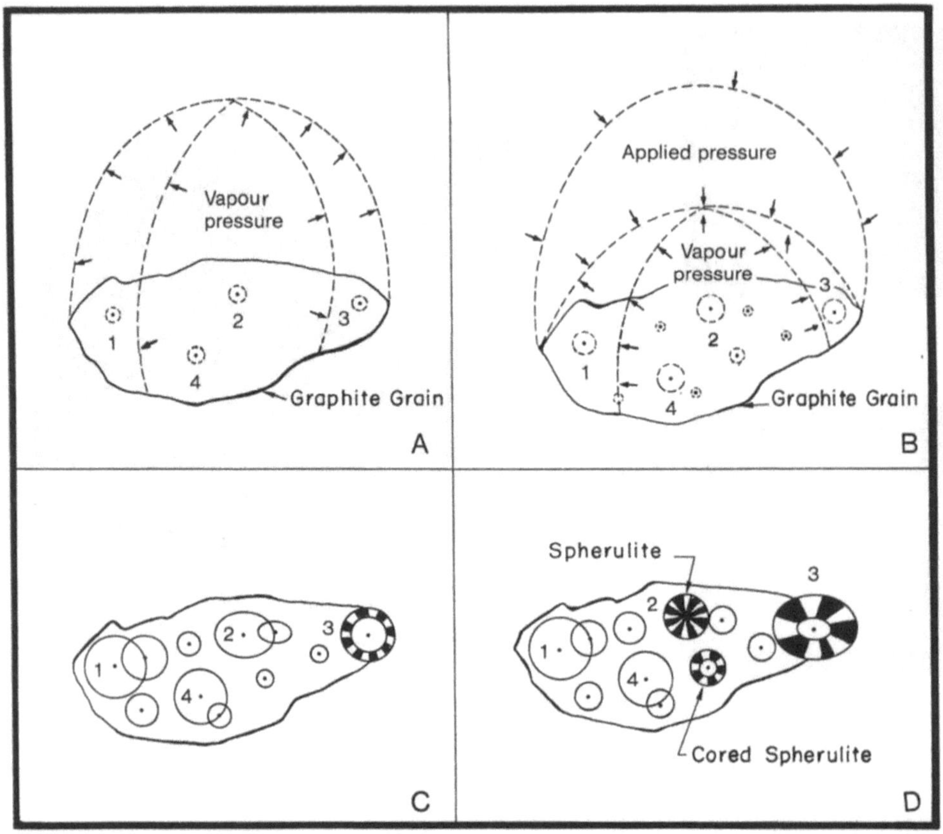
Vapour pressure
1
2
3
4
Graphite Grain
A
Applied pressure
Vapour pressure
1
2
3
4
Graphite Grain
B
1
2
3
4
C
Spherulite
1
2
3
4
Cored Spherulite
D

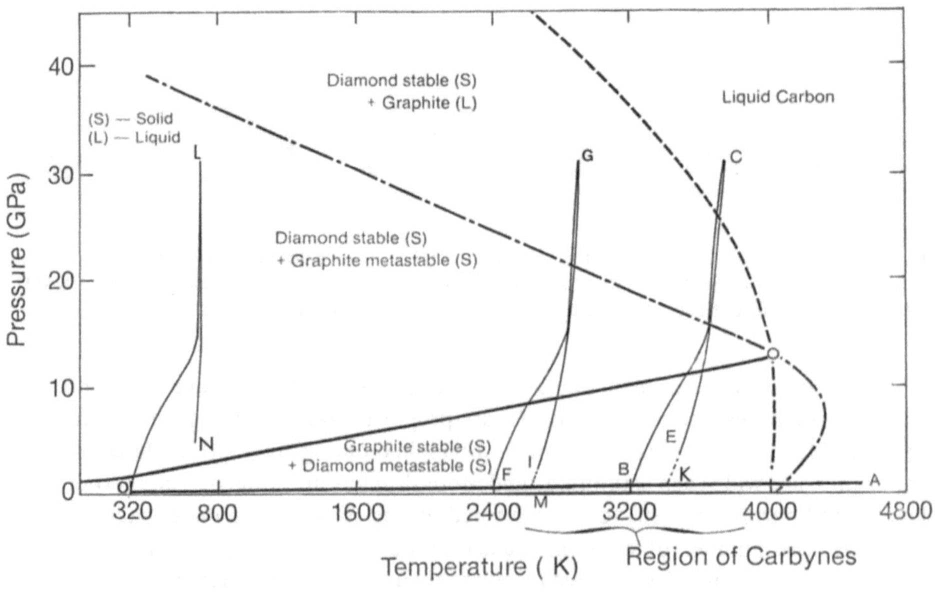
Diamond stable (S)
+ Graphite (L)
Liquid Carbon
(S) — Solid
(L) — Liquid
Diamond stable (S)
+ Graphite metastable (S)
Graphite stable (S)
+ Diamond metastable (S)
L
N
G
C
E
F
I
B
K
M
A
O
Pressure (GPa)
0
10
20
30
40
320
800
1600
2400
3200
4000
4800
Temperature (K)
Region of Carbynes

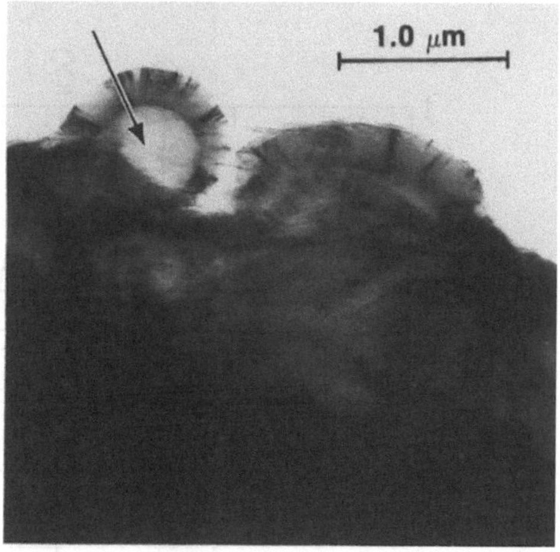

Fig. 38. Transmission electromicrograph of a partially regraphitized diamond grain, showing a rim consisting of a variety of other carbon polymorphs

case the sample cools down significantly before the pressure pulse is applied, and the activation energy supplied by the temperature pulse in turn is not sufficient to facilitate diamond nucleation (Fig. 37, path OLN).

These findings are consistent with the proposed model. In order to validate the model, a series of experiments has been performed whose statistical evaluation with a simple second-order model is shown in Fig. 39. It can be seen from the isopleths of yield of particles related to diamond formation that with increasing preheating energy and decreasing time interval between temperature and pressure pulses yield at constant pressure increases within the parameter limits investigated. The optimum time of offset of temperature and pressure pulses was determined to be around 100 μs at a pressure of 18 GPa[183].

5 Experimental Designs and Techniques of Growth of Diamond and Boron Nitride

5.1 General Remarks

The preservation of the newly formed dense phases of carbon or boron nitride in a shock experiment generally encounters difficulties caused by the specific character of shock compression of solids. Usually, as already discussed in the previous chapters, the interac-

◀ **Fig. 36A–D.** Kleiman-Heimann-Salansky model of formation of diamond during shock compression of graphite powder[5, 195]. **A)** Nucleation of liquid droplets in graphite (temperature high, pressure low). **B)** Nucleation of cubic diamond (temperature high, pressure high). **C)** Growth of nucleated particles (temperature high, pressure decreasing). **D)** Regraphization of diamond and recrystallization of carbon blacks (temperature decreasing, pressure low)

◀ **Fig. 37.** Superposition of three dynamic diagrams *(OLN, FGI, BCK)* with different preheating temperatures (320, 2400, 3200 K) on a phase-transition diagram of carbon[5]

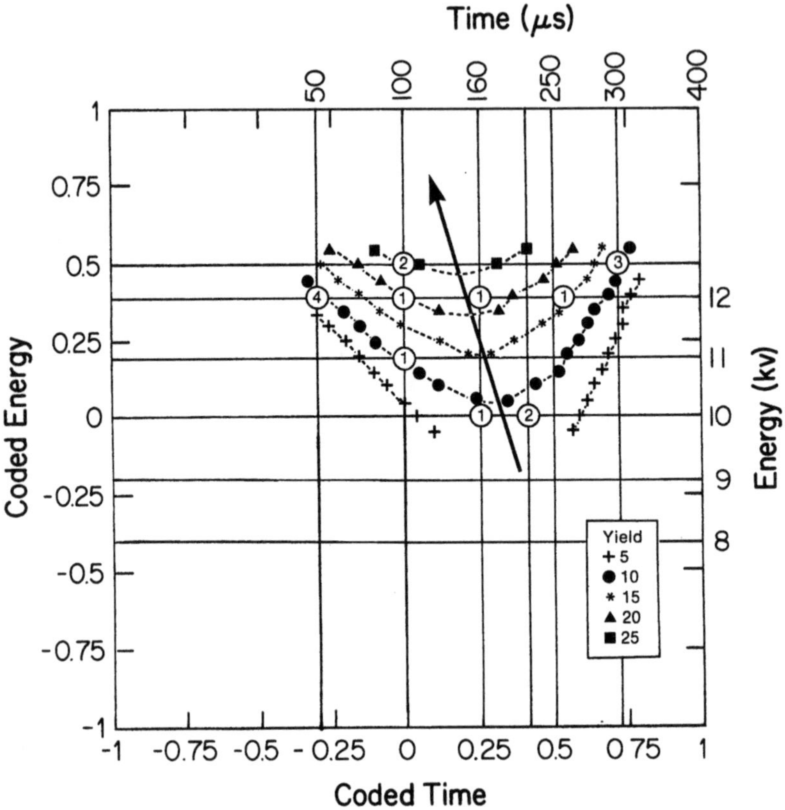

Fig. 39. Isopleths of yield of particles related to diamond formation (second-order model) fitted to experimental data by Kleiman et al.[5]. *Numbers in circles* represent the number of experiments conducted at particular combinations of discharge energy and time interval between temperature and pressure pulses. Pressure: 18 GPa, sample density: 75% of theoretical density

tion time with matter of the high-pressure pulse is by orders of magnitude shorter than the time during which the high temperature persists in the system. This situation can lead to extensive regraphitization of dense forms of carbon and boron nitride. To alleviate these drawbacks, special efforts were made to properly design the containers encapsulating the samples, and to effectively control the initial state of the sample, and the high pressure – and temperature – built up and release in the samples. Only then the relations between maximum pressure (P_{max}), maximum temperature (T_{max}) and residual temperature (T_{res}) can be controlled within a wide range, and conditions can be found where synthesis is accomplished and the high-pressure phases are preserved with a high yield. Tables 6 and 7 summarize the results obtained by several working groups around the world in shock synthesis of diamond and boron nitride. It should be mentioned that, in general, a characteristic feature of explosively synthesized diamonds is their ultradisperse size[5, 19, 36, 207, 208]. The shape of the diamonds depends on the mechanism of their formation. Diamonds synthesized by a solid-to-solid transformation mechanism (see Chap. 4), show typically angular crystals with a large number of cutting edges[36]. Spherically shaped diamonds seem to be formed if a liquid phase is involved[5, 188]. The surfaces of those particles are well developed and have a large number of pores[36, 183, 208]. The size of the synthesized diamonds depends upon the loading method used (see Table 3) and the grain size of the starting material. The shock synthesized diamonds possess highly defective structures. It was found by electron spin resonance measurements[208] that the concentration of free electrons, which tend to accumulate at lattice defects, can reach $(140–170) \times 10^{17}$ per gram. Dislocation densities in cubic diamond can reach values between 10^{12} cm^{-2} [209] and 10^{13} cm^{-2} [210]. The dislocation densities in lonsdaleite were found to be even higher by an order of magnitude than those reported for cubic diamond[210].

In the following sections some of the experimental set-ups will be described in more detail. It should be mentioned that despite the fact that the first announcement of successful diamond synthesis was made in the early sixties[1], a number of original papers and patents on potential diamond synthesis using different dynamic loading conditions already appeared in the late fifties[211, 186]. In addition to the work listed in Tables 6 and 7 a number of publications appeared during the last 25 years describing different devices aimed at dynamic shock synthesis of diamond[35, 212]. However, since no results in terms of actual diamond syntheses were published in these papers, they were excluded from the present review.

5.2 Diamond Synthesis

Following the announcement of successful shock-synthesis of diamonds in 1961[1], a number of designs for diamond shock synthesis were patented[33]. Table 8 shows some of the results obtained by DeCarli[33] using different experimental arrangements. A "mousetrap" planar shock-wave source in a direct-contact explosive configuration (Fig. 11) and a flying-plate loading arrangement were used. In the direct-contact method the high explosive rested on a block of carbon. In the flying-plate method, the high explosive rested on a flying plate which, in turn, rested on four supports. Upon detonation of the high explosive, a pressure pulse was applied to the carbon block and, in turn, to the supporting board. The board broke, thus quenching the shocked carbon in a tank with cold water. Different techniques can be used to retrieve the diamond particles formed and to separate them from the rest of the shocked, but unconverted material, and the debris. An arrangement similar to that described above but using a running detonation wave was applied directly to a block of high explosive in contact with a carbon block. As can be seen from Table 8 the yield of diamond varied from 1% to ~9% depending on the method used. The relatively low yields obtained[33] were attributed[2] to excessive regraphitization because of unfavorable pressure and temperature release histories in the carbon.

Using an arrangement similar to that described above, but introducing a metal cooling medium in close contact with the carbon, much higher yields were achieved (Table 6). It should be noted that the yield of diamond given in[2] was based on the amount of diamond in the amount of carbon recovered after the experiment rather than in the total amount of carbon used. Therefore, the reported yields[2] were somewhat higher. The yields given in Table 9 have been recalculated based on the total amount of carbon used. Both modifications of diamond were synthesized. Most particles, however,

Table 6. Summary of shock-induced synthesis of diamond

Initial sample						Experimental
Carbon (C)			Catalyst			
Type	Grain size (μm)	Density (Mg/m^3)	Type	Grain size (μm)	C/CAT.[f] (%)	Loading method
Artificial and natural graphite	N/A[a]	1.58–1.68	N/U[b]	N/U	N/U	Flyer plate
Pig iron (3% of graphite)	10–25	N/A	Pig iron	–	3/97	Flyer plate
Graphite (SP, Ceylon)			Fe Cu Al	20 20 5–10	5/95 3/97 9/91	
Natural graphite	2.5	0.8–1.34	Cu	30–200 (mesh)	8/92	Cylindrical flyer
Graphite	75	1.5–1.8	N/U	N/U	N/U	Hemispherical implosion
Technically pure graphite	N/A	1.7	Fe Cu	8–20	–	Direct contact explosive Flying plate Cylindrical flyer
"Non-diamond modification of carbon"[g]	N/A	N/A	N/A	N/A	N/A	Cylindrical flyer (?)
High-crystallinity natural graphite	4–6	N/A	N/U	N/U	N/U	Cylindrical flyer
Amorphous carbon: Lamp black, Detonation black	N/A	N/A	N/U	N/U	N/U	Cylindrical flyer and flash-heating
Graphite: SP-1 UCAR	2–10	1.6–2.05	Fe Cu	15–170 20–60	5–20/95–80	Hemispherical implosion and flash-heating
Graphite: (SP-1), JU electrode	25	0.9	Co	200	7/93	Two-Stage gas gun accelerated projectile
Glassy carbon	300 (mesh)	N/A	Cu	300 (mesh)	4/96	Flyer plate
$(CF)_n$	N/A	N/A	Cu	N/A	1:9 by weight	Flyer plate

[a] Information not available. [b] Catalyst not used. [c] Particles consist of 10 nm crystallites. [d] Particles consist of 10–20 nm crystallites. [e] Particles consist of 10–20 nm crystallites. [f] Ratio Carbon/Catalyst. [g] The type of material quoted from[3)]

were mixed structures of the cubic and the hexagonal forms. The hexagonal diamond was formed when (i) the graphite in the graphite-metal mixture was at high density (greater than 75% of theoretical), (ii) conditions for maximum cooling prevailed, and (iii) the amount of the metal cooling medium was up to 40% by volume of the mixture. Dependent on these conditions, up to 65% by weight of the diamond product was in the form of the hexagonal modification.

An even more dramatic increase in diamond yield was achieved when the planar flyer plate arrangement was substituted by a cylindrical one (Table 10). It was shown[31)] that

Conditions			Results			Ref.
Shock configuration	Pressure (GPa)	Temperature (K)	Type of diamond formed	Size (μm)	Yield (%)	
Planar	10–30	500–1300	Cubic	10[c]	2.6–9.3	1, 33
Planar	57–230	N/A	Cubic, Hexagonal	6–10	5.5–47	2
Cylindrical	40	2000 (°C)	Cubic, Hexagonal	–	42–76	31
Not clear	N/A	N/A	Cubic		3–5	213
Planar Planar Cylindrical	50–100	1800–2400	N/A	N/A	N/A	121
Cylindrical	7–13	4000–4500 (± 80)	Cubic	A few mμ[d]	N/A	3
Cylindrical	30–60	N/A	Cubic, Hexagonal	N/A	60 ± 10	188
Cylindrical	7–13	4000–4500	Cubic	N/A	N/A	116
Not clear	15–25	300–3200	Cubic	0.1–3.0[e]	5–10	5
Planar	20–40	N/A	Cubic	7 ± 3–18 ± 10 (nm)	0.05–2.5	19
Planar	100	N/A	Cubic	N/A	N/A	214
Planar	60	1270	Cubic	10 nm	20	215

the cylindrical driver technique when combined with addition of coolants to carbon gave better yields in carbon-to-diamond transformation. The improved diamond yields were attributed to the presence of near optimum pressures for diamond synthesis and to smooth pressure release.

Figure 40 shows schematically the arrangement used by Balchan and Cowan[31)]. A specimen (1) consisting of a carbon-metal mixture was enclosed in a cylindrical metal container (2). The ends of the container were closed by plugs (3, 4) and the whole assembly fitted coaxially inside another metal cylinder (5) which was used as a projectile.

Table 7. Summary of shock-induced synthesis of boron nitride

Initial sample						Experimental
BN			Catalyst			
Type	Grain size (μm)	Density (Mg/m³)	Type	Grain size (μm)	BN/CAT.[f]	Loading method
g-BN	N/A[a]	0.8–1.98	N/U[b]	N/U	N/U	Deceleration of explosive products Flyer plate
g-BN High purity & crystallinity	N/A	N/A	N/U	N/U	N/U	Cylindrical flyer (?) Flash heating
g-BN	1–5, 5–60	Porosity 7%	Cu	<300 mesh	5:95	Two-Stage gas gun accelerated projectile
g-BN	N/A	Porosity 9–20%	Cu	<300 mesh	5:95	Flyer plate
g-BN	1.5	N/A	Cu	150–300 mesh	5:95	Flyer plate
w-BN	0.1–0.5	N/A	Cu	150–300 mesh	5:95	Flyer plate
z-BN	1–5	N/A	Cu	150–300 mesh	5:95	Flyer plate
g-BN	N/A	2.19	N/U	N/U	N/U	Direct contact explosive
g-BN	N/A	N/A	N/A	N/A	N/A	Flyer plate (?)
r-BN	0.1–1	>1.89	Cu	200	4/96	Mousetrap generator
g-BN	3–5	0.3 · ϱ theor.	Al	30–90 mesh	3/97	Direct contact explosive
g-BN	N/A	N/A	N/U	N/U	N/U	Flyer plate

[a] Information not available. [b] Not used. [c] Particles consist of 10–20 nm crystallites. [d] Particles consist of 100 nm crystallites. [e] Only small amounts of z-BN were recovered. [f] Ratio boron nitride/catalyst

The distance between the cylinders was maintained by spacers (7, 8). A cylindrical layer of detonating sheet explosive (9) surrounded the projectile-cylinder (5). An electric blasting cap (11) initiated an explosive disc (10) which, in turn, initiated the cylindrical sheet explosive (9). A detonation front proceeded through the layer of sheet-explosive in an axial direction and caused the projectile-cylinder (5) to converge rapidly and to collide progressively with the lateral boundaries of the sample-containing cylinder (2).

The synthesized diamonds were in the form of coherent, fine-grained polycrystalline particles with a unique microstructure. This microstructure was characterized by a bimodal crystallite size distribution. Single particles containing very small crystallites, typically having diameters in the 1–4 nm range, were interspersed with much larger crystallites, that typically had diameters in the 10–16 nm range[31]. The electron diffraction patterns of the diamonds formed were characterized by diffuse diffraction rings upon which discrete diffraction spots were superimposed. The discrete spots were attributed to

Conditions			Results			Ref.
Shock configuration	Pressure (GPa)	Temperature (K)	Type of boron nitride formed	Size (μm)	Yield (%)	
Planar Planar	12–20 35–50	650–2100	w-BN & z-BN	N/A	0–80	16
Cylindrical	9–12	1000–3500	w-BN z-BN	1–2[c] 1.5[d]	25–40	216
Planar	55	N/A	w-BN	Conglomerates consisting of 20 nm crystallites		62
Planar	30–40	N/A	w-BN	N/A	N/A	217
Planar	60 100–200 60–200	N/A N/A N/A	– – –	– – –	70 N/A N/A	63
Cylindrical	14–30	N/A	w-BN & z-BN	μm sized particles	N/A	9
Planar (?)	20–70	N/A	w-BN & z-BN	N/A	1–50	158
Planar	40–100	N/A	w-BN & z-BN	N/A	N/A	69
Cylindrical	6	N/A	z-BN	N/A	N/A	31
Planar	47.5 (typical)		w-BN & z-BN[e]	50–100	~100	218

reflections from large crystallites, the rings were explained as reflections from smaller crystallites.

When an identical cylindrical sample holder was used with the explosive applied directly to its walls, the yield of synthesized diamonds was substantially reduced. This was due to the fact that only an extremely small portion of the sample was subjected to high pressure.

A few attempts were undertaken to use an additional source of heat during the compression of graphite or graphite/metal mixtures[5, 3, 118)]. In these works a flash-heating system attached to the high pressure facility was used to raise the temperature of the sample just prior to or during the application of the high pressure pulse. Kleiman et al.[5)] attached a flash-heating system to a hemispherical implosion chamber. The whole system was designed to create very strong independently controllable pressure and temperature pulses which can be applied to the sample at different times.

Table 8. Summary of shock synthesis conditions of diamond (from DeCarli[33])

Method	Initial sample			Pressure (GPa)		Temperature (K)		
	Material[b]	Density (Mg/m^3)	Mass (g)	Top	Bottom	Top	Bottom	Yield of diamonds (%)
Flying Plate	Carbon	1.58	340	N/A	N/A	N/A	N/A	2.35
	Carbon	1.68	1376	N/A	N/A	N/A	N/A	9.3
	Carbon	1.58	N/A[a]	30	16	1570	970	N/A
Direct Contact	Carbon	1.58	340	N/A	N/A	N/A	N/A	2.1
Explosive:								
Mousetrap Plane	Carbon	1.68	N/A[a]	24	15	1270	920	N/A
Generator	Carbon	1.68	N/A[a]	24	12	1270	820	N/A
see Fig. 11	Carbon	1.68	N/A[a]	16	10	970	770	N/A
Vertically Adjacent	Carbon	1.58	340	N/A	N/A	N/A	N/A	~1
	Carbon	1.68	N/A[a]	15	10	920	770	N/A
	Carbon	1.68	N/A[a]	15	8	920	670	N/A
Detonator	Carbon	1.68	N/A[a]	15	10	920	770	N/A

[a] The data in the examples described above were not sufficient to calculate the weight of the sample
[b] The type of carbon used in the described examples was not specified

Table 9. Summary of shock synthesis conditions of diamond (from Cowan et al.[2])

Sample: Type of material, Particle size (carbon/metal), Weight (carbon/metal) %	Yield of synthesized diamonds (%)	P (GPa)	Size of diam. part. (μm)
Pig iron with ~3% of graphite, (10–25 μm)	47	185	–
Pig iron with ~3% of graphite, (10–25 μm)	~35.5	230	8.2
Pig iron with ~3% of graphite, (10–25 μm)	24.2	97	6.8
Cast iron (3% of graphite with particle size 20–50 μm)	12	140	40% of particles >10 μm
Cast iron (3% of graphite with particle size 20–50 μm)	12	84	50% of particles ≧10 μm
Ceylon graphite/Fe, 30 μm/20 μm, (5/95)%	8.5	100	N/A
Spectroscopically pure graphite/Cu, 20 μm/20 μm, (2.7/97.3)%	11	95	N/A
Spectroscopically pure graphite/Al, 0.1 μm/5–10 μm, (8.5/91.5)%	5.5	57	N/A
Spectroscopically pure graphite/Ni, (2.7/97.3)%	14.3	98	N/A

Table 10. Summary of shock synthesis conditions of diamonds (from Balchan et al.[31])

Sample Type of material, Particle size (graph./metal), Weight (graph./metal) (%)	Applied pressure (GPa)	Yield (%)	Type of diamond[a] formed
Natural graphite/Cu 2.5 μm/74–105 μm (150–200 mesh) (8/92)%	40–41	75.9	Cubic, Lonsdaleite
Natural graphite/Cu 2.5 μm/60–80 mesh (8/92)%	40–41	62.9	Cubic, Lonsdaleite
Graphite/Cu 2.5 μm/30–40 mesh (8/92)%	40–41	42.7	Cubic, Lonsdaleite
Graphite/Cu 2.5/30–40 (mesh) (8/92)%	40–41	42.4	Cubic, Lonsdaleite
g-BN/Al 3.5 μm/30–90 mesh (3/97)%	16	–	z-BN and w-BN

[a] One experiment with BN was also reported in[31] and was added to this table

The implosion chamber used in[5] was an improved version of that utilized by Glass and co-workers[213, 219–221]. As shown in Fig. 41, the 20-cm diameter cavity of the hemispherical chamber rested on a massive supporting frame with its axis in a vertical position. Two hydraulic jacks (H) connected to a compressor could lift or lower the upper part of the chamber in about 60 s. Four studs (G) threaded at the ends, guided the upper part of the chamber to the lower plate. During the experiments, four large nuts (L) threaded on to the studs held the two parts of the chamber together in addition to the hydraulic jacks.

All manipulations of the vacuum and gas systems were performed remotely. Figure 42 shows the principal of operation of the chamber. During the experiment, the chamber was first filled with a stoichiometric mixture of hydrogen and oxygen to total a pressure of 2.8 MPa. The gases were then detonated by exploding a small nickel wire (0.1 mm diameter and 2 mm long) located at the centre of the hemisphere (Fig. 42(a)) through a capacitor discharge (1 μF × 20 kV). A detonation wave was created that propagated outward to the PETN explosive liner[220, 221]. The explosive was instantly and simultaneously detonated (Fig. 42(b)) upon reflection of the detonation wave and a strong, stable, focused implosion wave converged inwards and was reflected at the centre as a shock wave (Fig. 42(c)), leaving behind a high-temperature and high pressure region (Fig. 42(d)). The implosion also produced a complex shock-wave rarefaction-wave system in the sample holder containing the material. It should be noted that the idea of using a hemispherical or spherical implosion for exploration of phase-transitions was already expressed by DeCarli in 1966[33], followed by a similar design in 1972[35]. It was suggested in these patents to use a number of detonators to ignite a spherically-shaped explosive charge simultaneously, thus creating a very strong spherical implosion converging towards the sample located in the centre. However, it was found later[222] that the initial

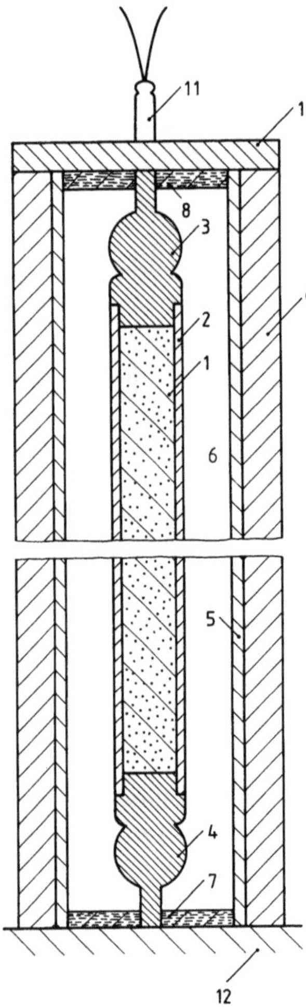

Fig. 40. Schematic representation of a cylindrical flying plate generator (after[31]). *1* – Graphite sample; *2* – Cylindrical metal container; *3,4* – Metal plugs; *5* – Metal tube-projectile; *6* – Air gap; *7,8* – Metal discs; *9* – High explosive; *10* – Explosive disc; *11* – Electric blasting cap; *12* – Metal anvil

"jetting" at the detonators completely destroyed the symmetry of the detonation of the hemispherical explosive pad. Instead of using a multitude of detonators[18], a single exploding wire was put at the geometrical centre of the hemisphere as described above. The flash-heating system consisted of a 160 μF capacitor bank connected in series to the sample by means of a spark gap and a low inductance/low-loss transmission line. The capacitors could be charged to a maximum 20 kV and held then an energy of 32 kJ. The capacitors could be discharged in 250 μs (90% of their energy was discharged within 50–100 μs[5]). The peak current through the sample could reach 100 kA at 20 kV. A wide range of P, T conditions was covered in experiments with the Flash-Heating Hemispherical Implosion System (FHHIS), (see Table 6). Diamonds with two distinct shapes were found[5]. When the high pressure implosion chamber was used in conjunction with the flash-heating systems the resulting diamond particles were usually spherical in shape. Table 11 summarizes the features found by transmission electron microscopy analyses on particles formed in flash-heating hemispherical implosion experiments (see Figs. 33–35).

18 91 detonators were used in[222]

Table 11. Summary of different features found in explosive-flash heating experiments (from Kleiman et al.[5])

1	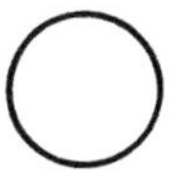	a) Round & irregular grains featuring diffraction patterns of cubic diamond and p-diamond. b) Round grains without any diffraction pattern.
2		Cored spherulites (Partially regraphitized spherulites with p-diamonds in center.)
3		Completely regraphitized spherulites.
4	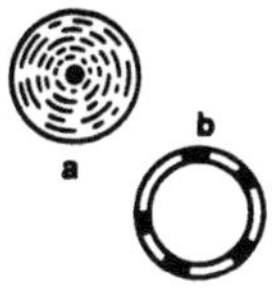	Round inclusions with a grain in the center and thin concentric layers (b) or radial rims (a) around the grains, featuring electron diffraction patterns of turbostratic graphite.
5		Angular inclusions arranged in conglomerates with turbostratic graphite diffraction patterns.
6		"Ghosts" (Particles featuring graphite electron diffraction patterns.)

When the experiments were conducted only with the hemispherical implosion chamber the diamond particles formed were irregular in shape, consisting of crystallites of 10–20 nm size fused into larger conglomerates. Figure 43 shows the shape of diamonds formed in experiments with the hemispherical implosion system, without the preheating system. The spherical shape of diamonds recovered in high-temperature shock experiments (see Fig. 33, Table 11)[5], (see also[188]) suggested a formation mechanism involving a liquid state, while the irregular angular shape of diamonds formed in shock experiments without flash-heating (Fig. 43) suggested a solid-solid transformation mechanism (see Chap. 4 for more details).

The size of diamonds obtained in a different investigation involving ultra-high temperatures, moderate pressures, and ultra-high cooling rates[3] were in the range of a few tens of nm. The majority of the crystals had a size of the order 10–20 nm. It was also found that the synthesized diamond crystals were single grains and practically undistorted. However, in the process of the high-temperature explosive synthesis, the separate crystals were found to be sintered together into floccular formations up to several μm in size.

Using high magnetic fields, typically in the range of 100 V · s/m^2, high pressures were transmitted to a hollow copper spool filled with graphite or a hollow aluminum spool filled with boron nitride[48]. Figure 44 shows schematically the arrangement used. A capacitor bank consisting of eight 15 μF × 20 kV capacitors arranged around two octagonal transmission plates *1, 2* and connected to them via a spark gap was designed to give up to 24 kJ of energy when discharged simultaneously through the transmission plates and the copper spool *3* attached to them. As shown in Chap. 3.3, a high pressure

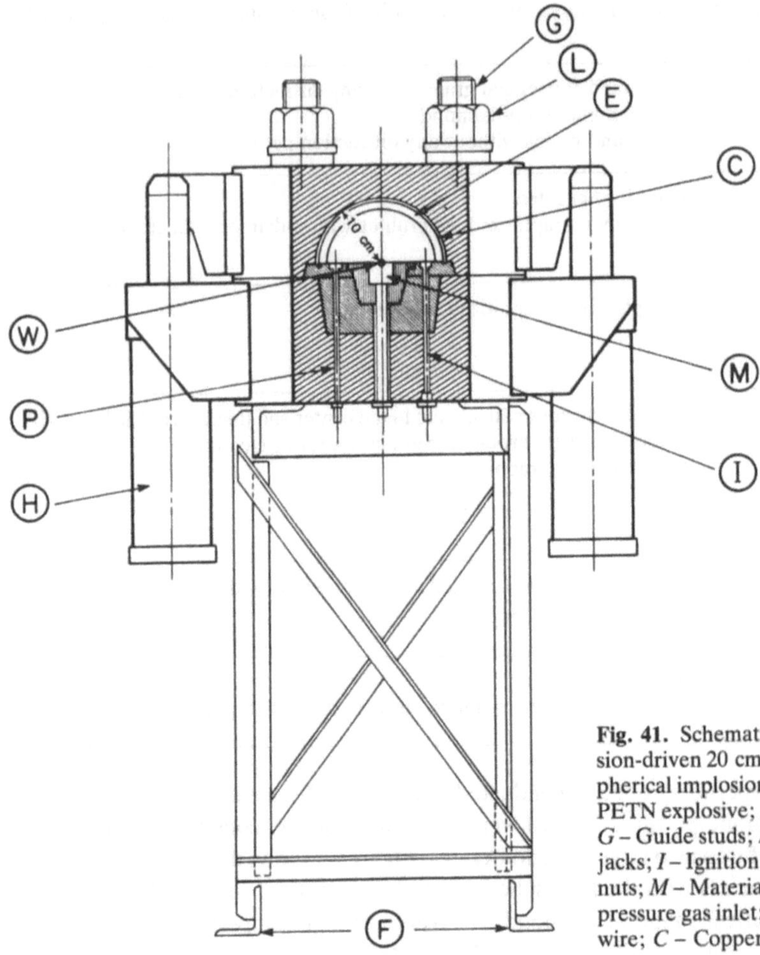

Fig. 41. Schematic view of an explosion-driven 20 cm diameter hemispherical implosion chamber. *E* – PETN explosive; *F* – Support frame; *G* – Guide studs; *H* – Hydraulic jacks; *I* – Ignition line; *L* – Lock nuts; *M* – Material holder; *P* – High-pressure gas inlet; *W* – Exploding wire; *C* – Copper liner

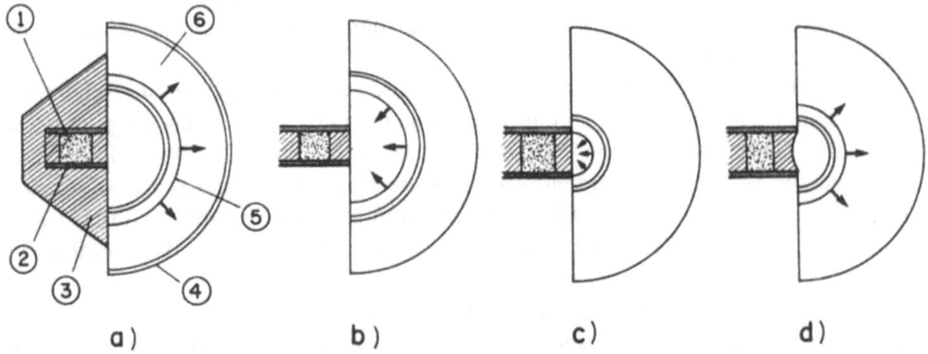

Fig. 42a–d. Schematic illustration of the principle of operation of a hemispherical implosion chamber. **(a)** Outgoing detonation wave of oxygen/hydrogen mixture: *(1)* sample material; *(2)* sample holder with plugs; *(3)* supporting crucible; *(4)* copper shell lined with explosive; *(5)* detonation front; *(6)* H_2–O_2 mixture. **(b)** Ignition of explosive material. **(c)** Converging shock wave. **(d)** Reflection of shock wave at upper plug surface

will result in the interior of the spool and will be transmitted eventually to the sample 4 contained by the spool. On applying a pressure of 18 GPa which was attained by using 7 capacitors charged to 17 kV, formation of cubic BN (z-BN) was achieved. Using pressures between 17–30 GPa formation of lonsdaleite was observed. In addition to the investigations discussed above a number of other attempts were undertaken to synthesize diamonds under shock conditions. Table 6 includes information on some of the results obtained in these works.

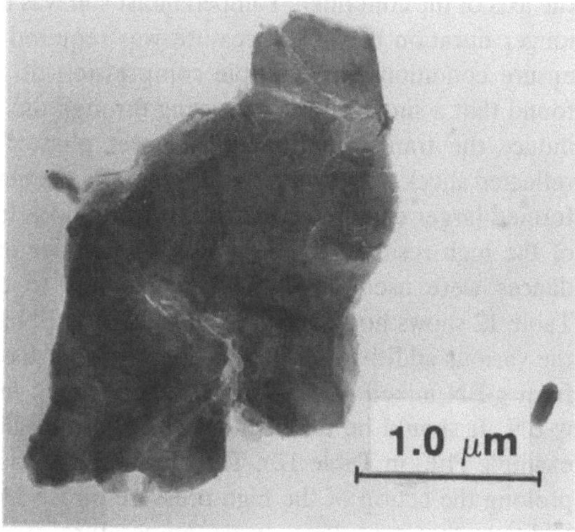

Fig. 43. Transmission electron micrograph of a diamond particle synthesized without the application of a flash-heating pulse[5)]

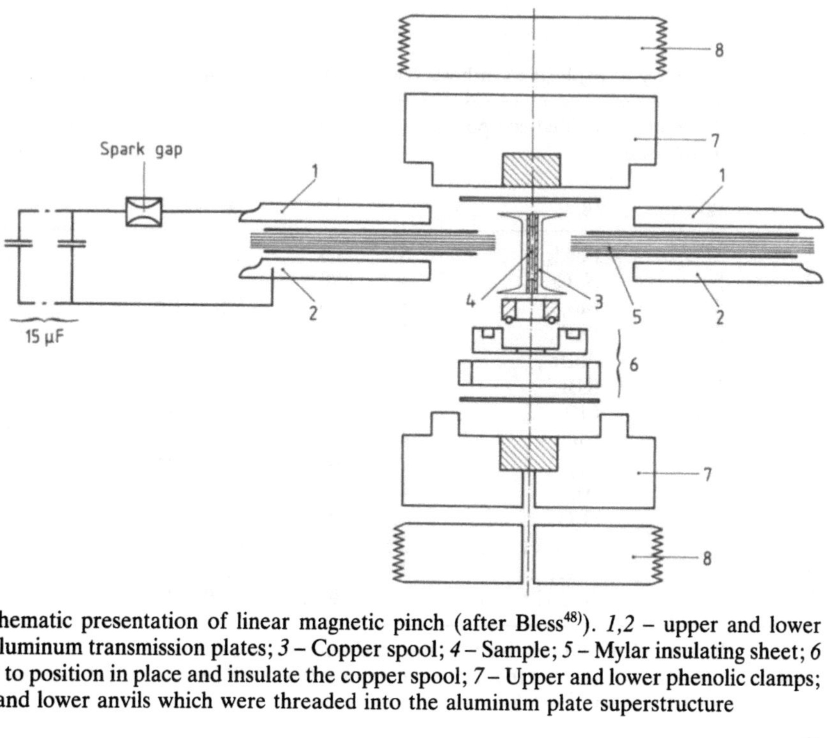

Fig. 44. Schematic presentation of linear magnetic pinch (after Bless[48)]). *1,2* – upper and lower octagonal aluminum transmission plates; *3* – Copper spool; *4* – Sample; *5* – Mylar insulating sheet; *6* – Assembly to position in place and insulate the copper spool; *7* – Upper and lower phenolic clamps; *8* – Upper and lower anvils which were threaded into the aluminum plate superstructure

5.3 Boron Nitride Synthesis

As can be seen from Table 7, considerable effort was put by many investigators into the shock synthesis of dense modifications of BN. All designs described in the previous section can be used, with some modifications, for the shock synthesis of z-BN and w-BN.

A cylindrical container was used to preserve the products of shock loading of g-BN[194, 223, 224]. In this arrangement, upon detonation of the explosive charge, the specimen enclosed in two cylindrical shells was compressed by a shock wave propagating along the axis of the container. Tamped moist soil was used around the explosive charges when longer duration of shock pressure was required[225]. Special precautions were taken to ensure conditions for multiple compression of g-BN in the container, because it was found that a single shock traversing through the graphitic boron nitride was not able to induce the transformation to a denser phase[224]. In case of multiple interactions of reflected shock waves with the BN sample, the newly formed particles of the dense phase formed larger conglomerates which were more resistant to regraphitization in the wake of the high residual temperatures. A number of materials with different shock impedances were used in mixtures with g-BN to enhance the effect of multi-shocking. Table 12 shows how sensitive the yield of w-BN is to differences in shock impedances of the various additives[194]. It was found[225] that the highest yields of w-BN were obtained from g-BN mixed with additives whose shock impedances were different from that of w-BN. It should be noted, however, that not all additives followed this trend (see for example PbF_2 in Table 12). The different materials added to g-BN act in two ways: they prolong the action of the high pressure on BN, and they play a role as coolants. Those additives that experience large temperature rises on shock wave interaction, can lower the residual temperature of the dense modifications of BN by melting, vaporizing or

Table 12. Summary of explosive synthesis conditions of BN (from Dremin et al.[194])

# of run	Amount of g-BN used (g)	Added compound (g)		Yield[a] (%)	Shock impedance of added[b] compound (at 12 GPa) ($g \cdot cm^{-2} \cdot s^{-1}$)	Density of formed w-BN (Mg/m^3)
1	5	$PbCl_2$,	5	60	$2.0 \cdot 10^6$	3.23
2	5	NaCl,	5	70	$1.1 \cdot 10^6$	3.28
3	5	Mo,	5	80	$5.5 \cdot 10^6$	3.38
4	2.24	PbF_2,	13.76	76	$3.0 \cdot 10^6$	3.31
5	2.8	$PbBr_2$,	17.2	80	$2.5 \cdot 10^6$	3.30
6	5.4	KBr,	5.4	50	$0.98 \cdot 10^6$	3.22
7	2.5	LiF,	2.5	60	$1.6 \cdot 10^6$	3.22
8	4	Cu,	4	50	$3.9 \cdot 10^6$	3.31
9	4	Fe,	4	60	$3.9 \cdot 10^6$	3.33
10	3	Al,	4	65	$1.7 \cdot 10^6$	3.32
11	2.5	H_2O,	2.5	95	$0.50 \cdot 10^6$	3.10
12	2.5	NaOH (1%),	2.5	99	$0.50 \cdot 10^6$	3.28
13	3	H_2O,	2	70	$0.50 \cdot 10^6$	3.20
14	2	NaOH (2%),	3	99.5	$0.50 \cdot 10^6$	3.25

[a] The yield was calculated relatively to the total amount of g-BN used

[b] Shock impedance of g-BN at the same conditions is $0.50 \cdot 10^6$ ($g \cdot cm^{-2} \cdot s^{-1}$) and that of w-BN is around $3 \cdot 10^6$ ($g \cdot cm^{-2} \cdot s^{-1}$)

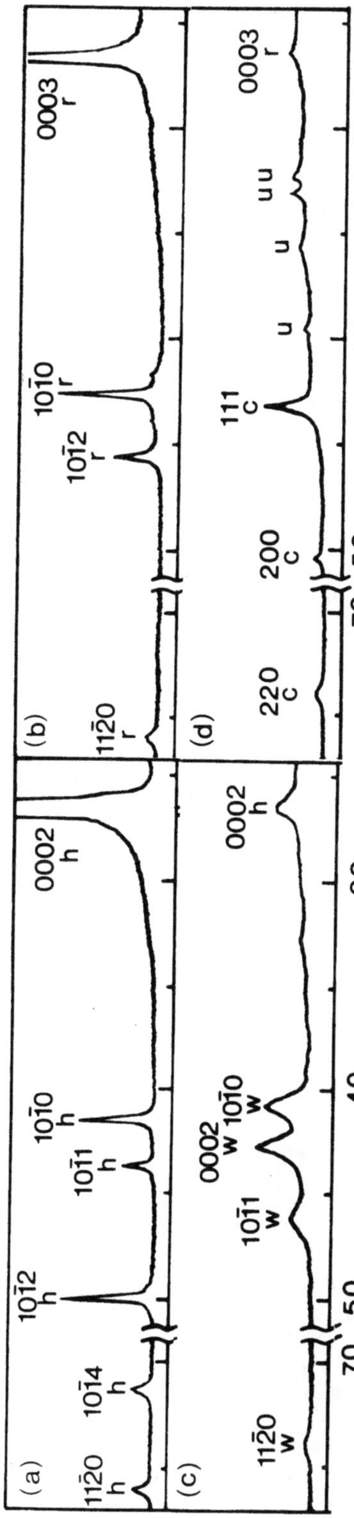

Fig. 45a–d. X-ray diffraction analysis of g-BN (**a, c**) and r-BN (**b, d**) before and after shock treatment. **h** – hexagonal, **r** – rhombohedral, **w** – wurtzite, **c** – cubic, **u** – unknown (courtesy N. Setaka[69])

decomposing around the BN formed. For instance, addition of NaCl can reduce the residual temperature of the mixture to 570 °C due to melting[225].

It was further found that the yield of the dense w-BN form was very low in areas adjacent to the center along the axis of the container, mainly because very high temperatures were developed leading to regraphitization. These shortcomings were overcome by inserting a metal rod through the center of the container[226]. Using a cylindrical container of this type, w-BN was synthesized from graphitic BN[227]. No traces of z-BN could initially be found in the reaction product. On close examination, however, z-BN was found to be present in the form of extremely fine lamellae randomly distributed throughout the w-BN grains and related to stacking faults. A characteristic feature of the reaction product was the presence of unreacted g-BN even after the various separation steps[228]. The residual g-BN has been attributed to factors such as unsufficient peak pressure during the impact leaving the material in the two-phase region on the Hugoniot curve, unfavourable orientation of g-BN, or regraphitization due to high residual temperatures[228].

Because of the controversy surrounding the formation of z-BN (see Chap. 4, 2.2)[16, 9, 217, 229] much effort was expended to better understand the conditions necessary for synthesis of the zincblende polymorph of BN[63, 69, 216, 229]. When graphitic boron nitride was subjected to a cylindrical shock wave, only ~1% of z-BN was recovered[229]. However, when a similar cylindrical container was used with a g-BN sample that was preheated, shocked and then very rapidly cooled, up to 20% of z-BN was found in the recovered sample[216]. At synthesis temperatures of only 2000 °C the product of transformation was pure w-BN. Only at temperatures around 3000 °C the z-BN phase was recovered. The diffusionless transformation g-BN → w-BN as observed by many researchers is thought to be the first stage of formation of z-BN. If sufficiently high temperatures are

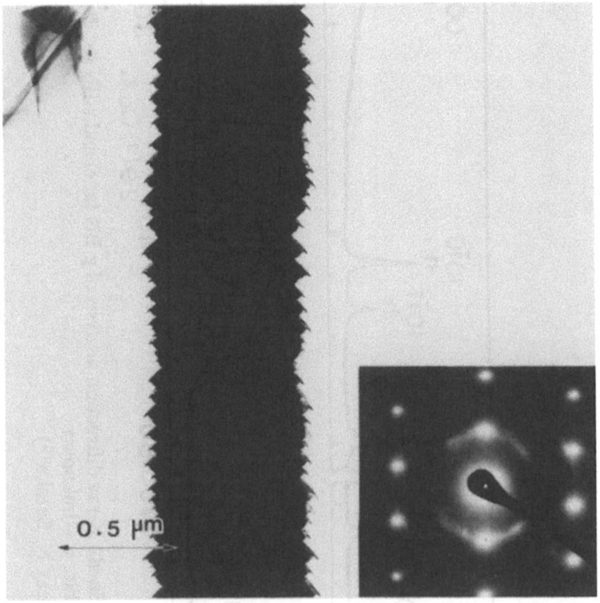

Fig. 46. Transmission electron micrograph of a z-BN crystal. The inset shows the electron diffraction pattern of the r-BN crystal (courtesy N. Setaka[69])

maintained, this stage is followed by the next stage of transformation of w-BN into z-BN via a diffusion process[63].

A different approach to synthesize BN was taken by Akashi et al.[63] who subjected different forms of BN to single and multiple shocking. A flying plate-planar shock wave generator was used to create high pressures in g-BN, w-BN and z-BN (see Table 7). The thickness of the flying plate was 3.2 mm. It was found that a single compression of g-BN yields only w-BN. Traces of z-BN were detected only after the sample was subjected to three consecutive shock loadings. In all cases the pressure was maintained at about 60 GPa. However, when w-BN was used as initial material, the transformation to z-BN was observed in a single shock experiment at pressure of 100 GPa. An increase of g-BN phase together with formation of z-BN was observed during double compression of w-BN using pressures of 150 GPa and 200 GPa, respectively. The increased amounts of g-BN were probably due to enhanced retransformation at higher pressures and, therefore, higher residual temperatures.

A different route was chosen by Sato et al.[69] to synthesize z-BN from its rhombohedral modification (r-BN) using a planar shock wave (see Table 7). Two sets of experiments were conducted where g-BN and r-BN were used as initial materials. It was shown that the g-BN sample was transformed into w-BN (Fig. 45(a, c)) while the r-BN form (Fig. 46) when subjected to pressures of 40–100 GPa yielded z-BN (Fig. 45(b, d)). Because the main diffraction peaks of Cu almost coincide with those of z-BN, caution was taken by the authors[69, 63] to keep the samples free of Cu. It was verified that the amount of copper was less than 0.1% wt[69] and less than 100 ppm[63], respectively, which in both cases was less than the detection limit in conventional X-ray analysis.

6 Conclusions

Successful phase transformation of the graphite-like forms of carbon and boron nitride into the high-pressure forms diamond and z-BN, respectively, has been carried out for the last twenty-five years. There is still, however, considerable controversy about the mechanism of the phase transition. One group of theories proposes the existence of a short-range-order structure, formed by collapse on interaction with the shock wave of the lattice of the precursor phase that becomes annealed at the prevailing high post-shock temperatures[91] and eventually crystallizes in the P-T stability field of diamond or z-BN. The "glassy" phase may be formed either via a quasi-liquid state in so-called "hot spots"[91] or by a solid state reaction involving global shearing displacement and catastrophical dislocation multiplication[177, 187].

On the other hand, the phase transition by shock wave interaction with matter may be explained as a diffusion-assisted reconstructive process[19] in which a very high density of mobile point defects is introduced by the shock wave sweeping through the crystal lattice[18]. These lattice defects give rise to a dramatic increase in the rate of diffusion of carbon atoms[18, 185]. Moreover, the shock causes stacking faults to spread and propagate involving climb and step formation[190]. The newly created steps in turn may coalesce and branch at partial-dislocation sources, and thus cause spreading stacking faults with c.c.p. layer arrangement to occupy different atomic planes.

Neither process, however, accounts for the nucleation of the high-pressure phase. Adiabatic shearing[90] by collision of micro-jets with cavity walls of collapsing pores[178] may create conditions that lead to true melting of the precursor phases. Furthermore, application of flash-heating pulses prior to the arrival of the shock wave can create temperatures sufficiently high to melt graphite[5, 174, 188]. These temperatures may in fact be so high that a substantial amount of carbon vaporizes instantaneously[202, 204], and subsequently forms droplets of liquid carbon by condensation of gaseous carbon in pore spaces[198]. This is at the heart of the SVLS model (solid-vapour-liquid-solid) proposed by Kleiman, Heimann and Salansky[5, 195].

There is some evidence that the actual process of nucleation and growth of diamond from graphite by shock loading encompasses solid state transformation as well as a transient liquid phase. Both mechanisms may coexist owing to highly inhomogeneous pressure and temperature distribution within a given sample[207, 215]. Similar conditions can exist in samples subjected to static pressures[230].

Whether the composite theory (see also[116]) is able to reconcile the discrepancies between the solid state transformation view and the liquid phase theory is unresolved to date. If the extremely high diffusion coefficients of carbon measured in shock compressed systems[20, 231] could be confirmed, then solid state transformation could be explained simply by movement of mobile point defects[19, 172] or by diffusion of carbon atoms between basal plans of graphite to create a chair-like configuration in c-direction[169]. Thus, phase transition could occur during the very short interaction time of the shock wave with the material to be transformed without the mediating action of a liquid phase.

Future development of experimental techniques could give direct information on lattice rearrangement thus complementing the pioneering work by Johnson and Mitchell[165] on in-situ X-ray diffraction studies of phase transformation in boron nitride employing an X-ray flash beam. Once a working X-ray laser system exists[232], it would give invaluable structural information on events whose duration is short compared with the time of existence of the shock-wave pulse.

Knowledge of the mechanism of the phase transition would enable the experimenter to optimize his or her experimental systems towards increasing the size of the diamond or z-BN particles obtained as well as increasing the yield. Up to the present, typical yields rarely exceed 60%, even in commercial shock-wave facilities. Cumbersome and time-consuming procedures to separate the high-pressure phases from the unreacted precursor phase and, if applicable, the metal powder would be greatly reduced if the yield could be substantially increased.

Manuscript received: January 9, 1986.

7 Notes Added in Proof

Considerable time has elapsed since submission of the manuscript of this review to the editor. In some areas covered in this paper new and exciting developments have occured that will be discussed briefly in the following "Notes added in proof".

A recent determination of the graphite/liquid/vapour triple point by measuring the spectral radiance of an extremely fast heated graphite sample yielded a value of 4530 +/− 150 K, 20 MPa[233]. This values is reasonably close to the values reported by Basset[142] and Gokcen et al.[143] (see 4.2.1).

2. There is indication that carbynes may be important precursor phases in the formation of diamond obtained by condensation of carbon from TNT detonation products[234]. The carbon phase change does not follow Bundy's equilibrium phase diagram[117] but favours apparently the stabilization of a low-density phase. This phase appears to have also a lower surface energy compared to diamond. Calculation of carbon cluster shapes and energies[235, 236] suggest that bare linear carbon chains, C_n, are stable against ring formation up to $n < 24$. Larger clusters can occur if chains condense, and reduce their surface energy by binding their active sites to oxygen or hydroxyl ions[234].

In this connection it is interesting to note that the only carbyne form found in nature, chaoite[237], has presumably a chain length of n = 11[147, 148] suggesting that this cluster size is particularly stable. From photoionization time-of-flight mass spectrometry studies of laser-vapourized graphite[235, 236] it was found that carbon clusters with (4n + 7) (n = 1 to 4) atoms have special stability. Chaoite is the carbyne form with the lowest possible "magic number" (n = 11). This phase has been identifies also as a reaction product of a laser beam with a graphite surface by Whittaker et al.[141].

The assignment of a linear structure for n = 11, however, has been disputed in a recent analysis by Van Vechten and Keszler[238]. These authors assume a kind of "paddle-wheel" structure for n = 11 whose chemical name is either tricyclo(3.3.3.0)undeca-2,7,9-triene-4,6,11-triyl ("TUTT") or tricyclo(3.3.3.0) undeca-2,7,9-triene ("TUT"). Clusters with the higher "magic numbers" n = 15, 19, and 23 ared being formed by successive addition of four carbon atoms to the graphite side sections of the (more stable) TUTT configuration.

These fused-ring structures, however, are likely to possess a high amount of intramolecular strain energy, and therefore ought to be considerable less stable than unstrained linear chains[235]. But then, similarly shaped molecules such as propella-hexaene have been synthesized recently[239].

3. In an attempt to model EOS data of multiphase solids with a liquid phase, Van Thiel et al.[240] used a scaling model for the liquid phase, and also considered electronic excitation states. This model describes well the graphite-diamond phase transition, i.e., the Berman-Simon line but deviations from the experimental results exist for the graphite melting line. Because the predictions do not take into account a sufficiently large volume change on melting of graphite, phase changes near the melting point, i.e., carbyne formation by breaking up of graphite basal planes as suggested by Whittaker[138] and Heiman et al.[147] could not be modeled. This, in effect, may be a reason why Van Thiel's model failed to match the shape of the experimental melting curve of graphite[241].

4. A patent has been issued to Akashi and Sawaoka[242] describing the dynamic compaction of composite materials containing diamond. Synthetic diamond powder is enclosed in a stainless steel ampoule, and an explosively-driven steel flyer plate is propelled against it at a velocity of up to 3 km/s, creating a strong shock pressure of some 100 GPa. The shock-sintered diamond is 85% as hard as natural single-crystalline diamond, and can be used for machining tool bits or in electronic devices requiring high thermal conductivity.

8 References

1. DeCarli, P. S., Jamieson, F. C.: Science *133,* 1821 (1961); Poulter Laboratories Tech. Rep. 001–61 (Feb. 8, 1961)
2. Cowan, G. R., Dunnington, B. W., Holtzmann, A. H. (To E. I. DuPont de Nemours & Co.): U.S. Patent 3,401,019 (1966)
3. Trefilov, V. I., Savvakin, G. I., Skorokhod, V. V., Solonin, Yu. M., Khrienko, A. F.: Dokl. Akad. Nauk SSSR *239,* 838 (1978)
4. Yamaguchi, S., Setaka, N.: Carbon *17,* 371 (1979)
5. Kleiman, J., Heimann, R. B., Hawken, D., Salansky, N. M.: J. Appl. Phys. 56(5), 1440 (1984)
6. Adadurov, G. A., Aliev, Z. G., Atovmyan, L. O., Bavina, T. V., Borod'ko, Yu. G., Breusov, O. N., Dremin, A. N., Muranevich, A. Kh., Pershin, S. V.: Dokl. Akad. Nauk SSSR *172,* 1066 (1967). [Sov. Phys.-Dokl. *12,* 173 (1967)]
7. Alekseevsky, V. P., Bochko, A. V., Dzhamarov, S. S. et al.: US Patent 3,876,751 (April 8, 1975)
8. Sawaoka, A., Akashi, T.: Mat Res. Soc. Symp. Proc. *24,* 365 (1984)
9. Coleburn, N. L., Forbes, J. W.: J. Chem. Phys. *48*(2), 555 (1968); Coleburn, N. L.: J. Chem. Phys. *40*(1) (1964)
10. Alder, B. J., Christian, R. H.: Phys. Rev. Letters *7,* 367 (1961)
11. Batsanov, S. S., Blokhina, G. E., Deribas, A. A.: Zh. Strukt. Khim. *6,* 227 (1965)
12. Duvall, G. E., Fowles, G. R.: in: High Pressure Physics and Chemistry (Bradley, R. S. ed.), Academic Press, New York *2,* 209 (1963)
13. Rosenberg, R. L., Alder, B. J.: Shocks up to Infinite Pressures, in: Metallurgy at High Pressures and High Temperatures, Metallurgical Society Conferences *22,* 302, Chap. 12; Gordon and Breach Science Publ./New York – London, 1963
14. Murri, W. J., Curran, D. R., Peterson, C. F., Crewdson, R. C.: Response of Solids to Shock Waves, in: Advances in High Pressure Research, Vol. 4 (Wentorf, R. H. ed.), London, New York, Academic Press, 1974
15. Samara, G. A.: in: High-Pressure Science and Technology, Proc. of 6th AIRAPT Conf., Boulder, CO (Timmerhaus, K. D., Barber, M. S. eds.), Plenum Press, NY and London *1,* 177 (1979)
16. Dulin, I. N., Al'tshuler, L. V., Vashchenko, V. Ya., Zubarev, V. N.: Fiz. Tverd. Tela *11*(5), 1252 (1969) [Sov. Phys. – Solid State *11*(5), 1026 (1969)]
17. Lonsdale, K., Milledge, N. J., Nave, E.: Min. Mag. *32,* 185 (1959)
18. Pujols, H., Boisard, F.: Carbon *8,* 781 (1970)
19. Morris, D. G.: J. Appl. Phys. *51,* 2059 (1980)
20. Alekseevskii, V. P., Mazanko, V. F., Pechentkovskaya, L. E., Falchenko, V. M.: Sov. Phys. – Tech. Phys. *24,* 477 (1979)
21. Heimann, R. B., Kleiman, J., Salansky, N. M.: J. Crystal Growth *67,* 213 (1984)
22. Alt'shuler, L. V.: Soviet Physics Uspechi *8(1),* 52 (1965)
23. Graham, R. A., Asay, J. R.: High Temperature – High Pressure *10,* 355 (1978)
24. Batsanov, S. S., Deribas, A. A., Kutolin, S. A.: Izvestiya AN SSSR, Neorg. Mat. *2,* 87 (1966)
25. Batsanov, S. S.: in: Dokl. Vses. Symp. Impulsnim Davleniam, 1-st, 1973, *2,* 4 (1974)
26. Rice, M. H., McQueen, R. G., Walsh, J. M.: Compression of Solids by Strong Shock Waves, in: Solid State Physics (Seitz, F., Turnbull, D. eds.), New York, Academic Press *6,* 1 (1958)
27. Walsh, J. M., Christian, R. H.: Phys. Rev. *97(6),* 1544 (1955)
28. Yampolskii, P. A.: Vestnik AN SSSR *4,* 42 (1975)
29. Zeldovich, Ya. B., Raiser, Yu. P.: Physics of Shock Waves and High-Temperature Hydrodynamic Phenomena, V. II, Academic Press, New York and London 1967
30. Isbell, W. M., Walton, O. R., Ree, F. H.: in: High-Pressure Science and Technology, Proc. 6th AIRAPT Conf., Boulder, CO. (Timmerhaus, K. D., Barber, M. S. eds.), Plenum Press, New York and London *1,* 68 (1979)
31. Balchan, A. S., Cowan, G. R.: US Patent 3,667,911 (June 1972)
32. Deal, W. E. Jr.: in: Modern Very High Pressure Techniques (Wentorf, R. H. ed.), Washington, DC, Butterworths, 240 (1982)

33. DeCarli, P.: U.S. Patent 3,238,019 (March 1966)
34. Fowles, G. R.: in: Dynamic Response of Materials to Intense Impulsive Loadings (Chou, P. C., Hopkins, A. K. eds.), Wright-Patterson Air Force Base, Ohio, Air Force Laboratory, 405 (1972)
35. Garrett, D. R.: U.S. Patent 3,659,972 (May 1972)
36. Andreev, V. D., Voloshin, M. N., Lukash, V. A., Sozin, Yu. I., Rizhikov, A.A.: Sverchtverdiye Materiali (Superhard Materials) *5,* 25 (1984)
37. Fowles, G. R., Duvall, G. E., Assay, J. R., Bellamy, P., Feistmann, F., Grady, D., Michaels, T., Mitchell, R.: Rev. Sci. Instrum. *41,* 984 (1970)
38. Graham, R. A.: J. Appl. Phys. *46,* 2153 (1971)
39. Barker, L. M., Hollenbach, R. E.: J. Appl. Phys. *45,* 4872 (1974)
40. Jones, A. H., Isbell, W. M., Maiden, C. J.: J. Appl. Phys. *37,* 3493 (1966)
41. Keeler, R. N., Royce, E. B.: Shock Waves in Condensed Media, in: Physics of High Energy Density (Caldirola, P., Knoepfel, H. eds.), Academic Press, New York and London, 1971
42. Ragan, C. E. III, Silbert, M. G., Diven, B. C.: J. Appl. Phys. *48,* 2860 (1977)
43. Shaner, J. W.: in: High Pressure Science and Technology, Proc. of the VIIth Intern'l. AIRAPT Conf., Le Creuset, France (Vodar, B., Marteau, Ph. eds.), *1,* 965 (1979)
44. Kidder, R. E.: Nuclear Fusion *8,* 3 (1968)
45. Kidder, R. E.: Interaction of Intense Photons and Electron Beams with Plasmas, in: Physics of High Energy Density (Caldirola, P., Knoepfel, H. eds.), Academic Press, New York and London, 1971
46. Holzrichter, J. F.: Nature *316,* 309 (1985), (see also related articles in this issue)
47. Knoepfel, H.: Cumulation of Electromagnetic Energy, in: Physics of High Energy Density (Caldirola, P., Knoepfel, H. eds.), Academic Press, New York and London, 1971
48. Bless, S. J.: J. Appl. Phys. *43,* 1580 (1972)
49. Bennet, F. D.: High Temperature Exploding Wires, in: Progress in High-Temperature Physics and Chemistry, V. 2 (Rouse, C. A. ed.), London, 1968
50. Bernal, J. D.: Proc. Roy. Soc. London [A] *106,* 749 (1924)
51. Lipson, H., Stokes, A. R.: Proc. Roy. Soc. London [A] 181, 101 (1942)
52. Holcombe, C. E.: USAEC Oak Ridge Y-12 Plant Report, Y-1887 (1973)
53. Bundy, F. P., Kasper, J. S.: J. Chem. Phys. *96,* 3437 (1967)
54. Frondel, C., Marvin, V. B.: Nature *214,* 589 (1967)
55. Donnay, G., Donnay, J. D. H.: 12th Int. Cong. Crystallography, Aug. 16–25, 1981, Ottawa, Ontario, Canada
56. Seager, A. F.: Min. Mag. *43,* 377 (1979)
57. Yamaguchi, S., Hasegawa, Y.-T.: Fresenius Z. Anal. Chem. *294,* 43 (1979)
58. Nat. Bur. Stand. (U.S.) Monogr. *20,* 22 (1983)
59. Ishii, T., Sato, T., Sekikawa, Y., Iwata, M.: J. Crystal Growth *52*(1), 285 (1981)
60. Harold, A., Marzluf, B., Perio, P.: Compt. rend. Acad. Sci., Ser. C *246*(12), 1866 (1958)
61. Meyer, H. O. A.: Am. Min. *70,* 344 (1985)
62. Soma, T., Sawaoka, A., Saito, S.: Mat. Res. Bull. *9,* 755 (1974)
63. Akashi, T., Sawaoka, A., Saito, S.: Report of the Research Laboratory of Engineering Materials *2,* 133 (1977)
64. Reynolds, W. N.: Physical Properties of Graphite, Amsterdam, London, New York: Elsevier, p. 2 (1968)
65. Wheeler, E. J., Lewis, D.: Mat. Res. Bull. *10,* 687 (1975)
66. Trueb, L. F.: J. Appl. Phys. *39,* 4707 (1968)
67. Anders, E.: Astrophys. J. *134,* 1006 (1961)
68. Vereshchagin, L. F., Kalashnikov, Ya. A., Feklichev, E. M.: Dokl. Akad. Nauk SSSR *162,* 1027 (1965) [Sov. Phys. Dokl. *10,* 534 (1966)]
69. Sato, T., Ishii, T., Setaka, N.: J. Am. Ceram. Soc. *65*(10), C-162 (1982)
70. Milledge, H. J., Nave, E., Weller, F. H.: Nature *184* (4687), 715 (1959)
71. Nikol'skaya, I. V., Prezman, L. M., Zorkii, N. M. et al.: Zh. Fiz. Khim. *43,* 2948 (1969)
72. Riter, J. R.: J. Chem. Phys. *52,* 5008 (1970); J. Chem. Phys. *59,* 1538 (1973)
73. Kurdyumov, A. V.: Dokl. Akad. Nauk. SSSR *221*(2), 322 (1975) [Sov. Phys. Dokl. *20*(3), 218 (1975)]
74. Bundy, F. P., Strong, H. M., Wentorf, R. H.: in: Chemistry and Physics of Carbon, Vol. 10, p. 213 (1971), Marcel Dekker, New York (Walker Jr., P. L., Thrower, P. A. eds.)

75. Kurdyumov, A. V.: Fiz. Tverd, Tela *17,* 2469 (1975) [Sov. Phys. Solid State *17*(8), 1641 (1976)]
76. Estrin, E. I.: Fiz. Met. Metalloved *37,* 1249 (1974)
77. Bundy, F. P., Wentorf, R. H.: J. Chem. Phys. *38,* 1144 (1963)
78. Corrigan, F. R., Bundy, F. P.: J. Chem. Phys. *63,* 3812 (1975)
79. Bacon, G. E.: Acta Cryst. *3,* 320 (1950)
80. Boehm, H. P., Hofmann, U.: Z. anorg. allg. Chem. *278,* 58 (1955)
81. Laves, F., Baskin, Y.: Z. Krist. *107,* 337 (1956)
82. Freise, E. J., Kelly, A.: Phil. Mag. *8,* 1519 (1963)
83. Kochanovska, A.: Czech. J. Phys. *3,* 193 (1953)
84. Matuyama, E.: J. Phys. Chem. *58,* 215 (1954)
85. Freise, E. J., Kelly, A.: Proc. Roy. Soc. A *264,* 269 (1961)
86. Amelinckx, G.: Disc. Faraday Sec. *38,* 7 (1964)
87. Sclar, C. B., Squiller, S. F.: Meteoritics *11,* 360 (1960)
88. Jagodzinski, H.: Acta Cryst. *2,* 298 (1949)
89. Amelinckx, S., Delavignette, P.: J. Appl. Phys. *31,* 2126 (1960)
Delavignette, P., Amelinckx, S.: J. Nucl. Mater. *5,* 17 (1962)
90. Blackburn, J. H., Seely, L. B.: Nature *194,* 370 (1962)
91. DeCarli, P. S.: in: Science and Technology of Industrial Diamonds, Vol. 1 (Industrial Diamond Information Bureau, London), 49 (1967)
92. Neuhaus, A.: Angew. Chemie *66,* 525 (1954)
93. Bundy, F. P., Hall, H. T., Strong, H. M., Wentorf, R. H.: Nature *176,* 51 (1955)
94. Roozeboom, B.: in: Die heterogenen Gleichgewichte, p. 180 (1901), Vieweg, Braunschweig
95. Ludwig, A.: Z. Elektrochem. *8,* 273 (1902)
96. Parsons, C. A.: Proc. Roy. Soc. (London) *79,* 523 (1907)
97. Tammann, G.: Z. Phys. Chem. *69,* 569 (1909)
98. Moissan, H.: Comptes rend. Acad. Sci. *118,* 320 (1894)
99. Gunther, P. L., Geselle, P., Rebentisch, W.: Z. anorg. allg. Chemie *250,* 357 (1943)
100. Bridgman, P. W.: J. Chem. Phys. *15,* 92 (1947)
101. Burton, C. V.: Nature *72,* 397 (1905)
102. Sebba, F., Sugarman, N.: Nature *316,* 220 (1985)
103. Rossini, F. D., Jessup, R. S.: J. Res. Natl. Bur. Stand. U.S. *21,* 491 (1938)
104. Leipunskii, O. I.: Usphek Khim. *8,* 1519 (1939)
105. Berman, R., Simon, F.: Z. Elektrochem. *59,* 333 (1955)
106. Bundy, F. P., Bovenkerk, H. P., Strong, H. M., Wentorf, R. H.: J. Chem. Phys. *35,* 383 (1961)
107. Aisenberg, S., Chabot, R.: J. Appl. Phys. *42,* 2953 (1971)
108. Sokolowski, M., Sokolowska, A., Gokieli, B., Michalski, A., Rusek, A., Romanowski, Z.: J. Crystal. Growth *47,* 421 (1979)
109. Vora, H., Moravec, T. J.: J. Appl. Phys. *52,* 6151 (1981)
Matsumoto, S., Matsui, Y.: J. Mat. Science *18,* 1785 (1983)
Mori, T., Namba, Y.: J. Appl. Phys. *55(9),* 3276 (1984)
110. Eversole, W. G.: Canadian Patent 628,567 (1961); U.S. Patent 3,030, 187–8 (1962)
111. Derjagin, B. V., Fedoseev, D. V.: Sci. Amer. *233,* 102 (1975)
112. Kamo, M., Sato, Y., Matsumoto, S., Setaka, N.: J. Crystal Growth *62,* 642 (1983)
113. Margave, J. L., Bautista, R. G., Ficalora, P. J., Badachhape, R. B.: U.S. Patent 3,711, 595 (1973)
114. DeBeers Ind. Diamond Div. Ltd. (Johannesburg, S. A.), Austrian Patent 340,364 (1977)
115. Kennedy, C. S., Kennedy, G. C.: J. Geophys. Research *81,* 2467 (1976)
116. Trefilov, V. I., Savvakin, G. I.: Sov. Phys. – Doklady. *24,* 489 (1979)
117. Bundy, F. P.: J. Geophys. Research *85* (B12), 6930 (1980)
118. Bundy, F. P.: J. Chem. Phys. *38,* 631 (1963)
119. Fateeva, N. S.,Vereshchagin, L. F.: Pisma Zh. Eksp. Teor. Fiz. *13,* 157 (1971)
120. Bundy, F. P.: J. Chem. Phys. *38,* 618 (1963)
121. Deribas, A. A., Staver, A. M.: Fizika Goreniya i Vzryva, 13, 477 (1977) [English translation: Combustion, Explosion and Shock Waves *13,* 410 (1977)]
122. van Vechten, J. A.: Phys. Rev. *B7,* 1479 (1973)
123. Bundy, F. P.: J. Chem. Phys. *41,* 3809 (1964)

124. Jamieson, J. C.: Science *139,* 762 (1963)
125. Bundy, F. P.: Proc. Acad. Wetenschappen Amsterdam *72*(B), 302 (1969)
126. Mao, H. K., Bell, P. M.: Science *200,* 1145 (1978)
127. Grover, R.: J. Chem. Phys. *71*(9), 3824 (1979)
128. Pavlovskii, M. N.: Sov. Phys. – Solid State *13,* 741 (1971)
129. McSkimin, H. J., Andreatch, P.: J. Appl. Phys. *43,* 2944 (1972)
130. Bassett, W. A., Weathers, M. S., Gold, J. S., Bird, J. M.: Eos *64,* 847 (1983)
131. Gold, J. S., Bassett, W. A., Weathers, M. S., Bird, J. M.: Science *225,* 921 (1984)
132. Vereshchagin, L. F., Yakovlev, E. N., Stepanov, G. N., Vinogradov, B. V.: Pisma Zh. Eksp. Teor. Fiz. *16,* 382 (1972)
133. Kawai, N.: Junkatsu *21,* 475 (1976)
134. Kondo, K., Ahrens, T. J.: Geophys. Res. Letters *10*(4), 281 (1983)
135. Gust, W. H.: Phys. Rev. B *22*(10), 4744 (1980)
136. Derjagin, B. V., Fedoseev, D. V., Varnin, V. P., Vnukov, S. P.: Nature *269,* 398 (1977)
137. Kasatochkin, V. I., Savranskii, V. V., Smirnov, B. N. et al.: Dokl. Akad. Nauk SSSR *217*(4), 796 (1974)
138. Whittaker, A. G.: Science *200,* 763 (1978)
139. Whittaker, A. G., Kintner, P. L.: Carbon *23*(3), 255 (1984)
140. Whittaker, A. G., Wolten, G. M.: Science *178,* 54 (1972)
141. Whittaker, A. G., Kintner, P. L., Nelson, L. S., Richardson, N.: Aerospace Corporation Report SD-TR-81-60, NTIS No. A17106134, Sept. 1/81
142. Basset, M. J.: J. Phys. Radium *10,* 217 (1939)
143. Gokcen, N. A., Chang, E. T., Poston, T. M., Spencer, D. J.: High Temp. Sci. *8,* 81 (1976)
144. Whittaker, A. G.: Carbon *17,* 21 (1979)
145. Whittaker, A. G., Kintner, P. L.: Personal Communication
146. Whittaker, A. G.: Science *229,* 485 (1985)
147. Heimann, R. B., Kleiman, J., Salansky, N. M.: Nature *306,* 164 (1983)
148. Heimann, R. B., Kleiman, J., Salansky, N. M.: Carbon *22,* 147 (1984)
149. Haymet, A. D. J.: Chem. Phys. Letters *122*(5), 421 (1985)
150. Smith, P. P. K., Buseck, P. R.: Science *216,* 984 (1982)
151. Smith, P. P. K., Buseck, P. R.: Science *229,* 486 (1985)
152. Sladkov, A. M.: Sov. Sci. Rev. Sect. *B3,* 75 (1981)
153. Melnitchenko, V. M., Nikulin, Yu. N., Sladkov, A. M.: Carbon *23*(1), 3 (1985)
154. Franklin, R. E.: Proc. Roy. Soc. *209,* N1097, 196 (1951)
155. Noda, T., Inagaki, M.: Bull. Chem. Soc. Japan *37,* 1534 (1964)
156. Whittaker, A. G., Tooper, B.: J. Am. Ceram. Soc. *57*(10), 443 (1974)
157. Wentorf, R. H.: J. Chem. Phys. *26,* 956 (1957)
158. DeCarli, P. S.: Bull. Am. Phys. Soc. *23,* 1127 (1967)
159. Alt'shuler, L. V., Pavlovskii, N. M., Drakin, V. P.: Zh. Eksp. Teor. Fiz. *52,* 400 (1967) [Sov. Phys. – JETP *25,* 260 (1967)]
160. Sawaoka, A., Soma, T., Saito, S.: Japan J. Appl. Phys. *13,* 891 (1974)
161. Gust, W. H., Young, D. A.: Phys. Rev. B *15*(10), 5012 (1977)
162. Susa, K., Kobayashi, T., Tanaguchi, S.: Mat. Res. Bull. *9,* 1443 (1974)
163. Tani, E., Soma, T., Sawaoka, A., Saito, S.: Japan J. Appl. Phys. *14*(10), 1605 (1975)
164. Kurdyumov, A. V., Pilyankevich, A. N. O.: Phazovije Prevrashchenia v Uglerode i Nitride Bora [Phase Transformation in Carbon and Boron Nitride], Naukova Dumka, Kiev, 1979
165. Johnson, Q., Mitchell, A. C.: Phys. Rev. Letters *29*(20), 1369 (1972)
166. Grady, D. E.: J. Geophys. Res. B *85*(2), 913 (1980)
167. Lipschutz, M. E.: Science *45,* 1047 (1964)
168. Coulson, C. A., Haigh, C. W.: Tetrahedron *19,* 527 (1963)
169. Chomenko, A. A., Gankevich, L. T., Kotonosov, A. S., Kuteinikov, A. F.: Sint. Almazy *3,* 3 (1975)
170. Maire, J., Mering, J.: Chem. Phys. Carbon *6,* 125 (1970)
171. Coulson, C. A., Senent, S., Herraez, M. A., Leal, M., Santos, E.: Carbon *3,* 445 (1966)
172. Mogilevskii, M. A.: 2nd Vsesoyuz. Simp. Gor. i. Vzryvu, Akad. Nauk Armyansk. SSR, Chernogolovka, Oct. 25–30, 1969, p. 281 (1969)
173. Damask, A., Dins, D.: Tochechnye Defekty v Metallakh [Point Defects in Metals], Moscow, Mir, 1966

174. DeCarli, P. S.: in: High Pressure Science and Technology (Timmerhaus, K. D., Barber, M. S. eds.), 6th AIRAPT Conf., Boulder, CO., 1977; (Plenum, New York) *1,* 940 (1979)
175. McQueen, R. G., Marsh, S. P.: in: Behavior of Dense Media Under High Dynamic Pressure, Gordon and Breach, New York, 1968, p. 207
176. Duvall, G. E., Graham, R. A.: Rev. Mod. Phys. *49,* 523 (1977)
177. Horie, Y.: Phys. Rev. B *21*(12), 5549 (1980)
178. Seely, L. B.: Proc. 4th Electric Initiator Symp., Franklin Inst., Philadelphia, PA, 1963
179. Nesterenko, V. F.: High Pressure in Science and Technology, Mat. Res. Soc. Symp. Proc. (Homan, C., MacCrone, R. K., Whalley, E. eds.), (North-Holland, NY) *22,* 195 (1984)
180. Wackerle, J., Johnson, J., Halleck, P.: Proc. 6th Int. Symp. of Detonation, Coronado, CA, Aug. 24–27, 1976
181. Schilperoord, A.: Proc. 6th Int. Symp. on Detonation, Coronado, CA, August 24–27, 1976
182. Vereshschagin, L. F., Yakovlev, E. N., Buchnev, L. M., Dymov, B. K.: Teplofiz. Vysok, Temp. *15,* 316 (1977)
183. Kleiman, J.: Ph. D. Thesis, University of Toronto, 1983
184. Galimov, E. M.: Nature *243,* 389 (1973)
185. Adadurov, G. A., Shekhtman, V. Sh.: Vtor. Vses. Simp. Gor. i Vzryvu, Akad. Nauk. Armyansk. SSR, Chernogolovka, Oct. 25–30, 1969, p. 275 (1969)
186. Riabinin, Iu. N.: Zhur. Tekh. Fiz. *26,* 226 (1956)
187. Thompson, J. H. C.: Philos. Mag. *44,* 131 (1953)
188. Balan, T. R., Borimchuk, N. I., Bochko, A. B., Kurdiumov, A. V., Ostrovskaja, N. F., Piliankevich, A. N., Sedinkin, V. A., Francevich, I. N., Yarosh, U. V.: Sverchtverdiye Materialy [Superhard Materials] *3,* 19 (1983)
189. Roscoe, C., Thomas, J. M.: Carbon *4,* 383 (1966)
190. Solov'ev, V. A.: High Temp. High Press. *8,* 756 (1976)
191. Breusov, O. N.: Dokl. Vses. Simp. Impul'snyum Davleniam *2,* 18 (1974)
192. Breusov, O. N.: Vtor. Vses. Simp. Gor. i Vzryvu, Akad. Nauk. Armyansk. SSR, Chernogolovka, Oct. 25–30, 1969, p. 289 (1969)
193. Soma, T., Sawaoka, A., Saito, S.: Proc. 4th Int. Conf. High Pressure (Jiro, O. ed.), p. 446 (1975)
194. Dremin, A. N., Breusov, O. N., Bavina, T. V., Pershin, S. V.: DDR Patent 97638 (12 May, 1973)
195. Kleiman, J., Salansky, N. M., Heimann, R. B.: High Pressure in Science and Technology, Mat. Res. Soc. Symp. Proc. (Homan, C., MacCrone, R. K., Whalley, E. eds.), (North-Holland, NY) *22,* 223 (1984)
196. Kleiman, J., Salansky, N. M., Heimann, R. B.: J. Mat. Sci. Letters *3,* 117 (1984)
197. Trueb, L. F.: J. Appl. Phys. *41,* 5029 (1970)
198. Washi, F. M., Gilmartin, E. I.: Tech. Rept. TR-0200 (4250-40)-06, Vol. 1, Aerospace Corp., CA, Jan. 27, 1969
199. Heckman, F. A., Harling, D. F.: Rubber Chem. Technol. *59,* 1 (1966)
200. Lahaye, J., Prado, G.: in: Particulate Carbon Formation During Combustion (Siegle, D. C., Smith, G. W. eds.), Plenum, New York, 1981, p. 33
201. Donnet, J. B., Voet, A.: Carbon Black, Dekker, New York, 1976
202. Medalia, A. I., Rivin, D.: in: Characterization of Powder Surfaces (Parfitt, G. D., Singh, K. S. W. eds.), Academic Press, London – New York – San Francisco, 1976, p. 279
203. Salansky, N. M., Mar, H., Hawken, D., Kleiman, J.: Abstract of the Shock Wave Symp., Palo Alto, CA, April 1981 (unpublished); Kleiman, J., Salansky, N. M.: Phys. Canada *37* (N3), 41 (1981)
204. Abrahamson, J.: Carbon *11,* 337 (1973)
205. Girifalco, L. A., Lad, R. A.: J. Chem. Phys. *25,* 693 (1956)
206. Whittaker, A. G., Kintner, P.: Carbon *7,* 717 (1969)
207. Staver, A. M., Gubareva, N. N., Lyamkin, A. I., Petrov, E. A.: Combustion, Explosion and Shock Waves *20,* 567 (1984), (Translated from Fizika Goreniya i Vzryva *20,* 100 (1984))
208. Trueb, L. F.: J. Appl. Phys. *42,* 503 (1971)
209. Beliankina, A. V., Sozin, Yu. I., Tovstogan, V. M.: Sverchtverdiye Materiali [Superhard Materials] *4,* 18 (1981)
210. Sozin, Yu. I., Beliankina, A. V., Svirid, A. A.: Sverchtverdiye Materiali [Superhard Materials] *1,* 16 (1983)

211. Tilburg, J. V.: Australian Patent 226,337 (Nov. 1959)
212. Rasquin, J. R., Estes, M. F.: US Patent 3,632,242 (1972)
213. Glass, I. I., Sagie, D.: Phys. Fluids *25,* 269 (1982)
214. Setaka, N., Sekikawa, Y.: J. Mat. Sci. Lett. *16,* 1728 (1981)
215. Yamaguchi, S., Setaka, N.: J. Electrochem. Soc. *127,* 245 (1980)
216. Kurdyumov, A. V., Ostrovskaya, N. F., Pilipenko, V. A., Pilyankevich, A. N., Savvakin, G. I., Trefilov, V. I.: Sov. Phys. Dokl. *24,* 478 (1979)
217. Sawaoka, A., Soma, T., Saito, S.: Jap. J. Appl. Phys. *13,* 891 (1974)
218. Corrigan, F. R.: U.S. Patent 4,201,757 (May 1980)
219. Flagg, R. F., Glass, I. I.: Phys. Fluids *11,* 2282 (1968)
220. Glass, I. I.: Progress in Aerospace Science *13,* 223 (1972)
221. Glass, I. I., Sharma, S. P.: AIAA J. *14,* 402 (1976)
222. Kennedy, J. E., Glass, I. I.: Technical Note *99,* Univ. of Toronto, Institute for Aerospace Studies (1966)
223. Adadurov, G. A. et al.: Sov. Phys. Dokl. *12,* 173 (1967)
224. Adadurov, G. A., Bavina, T. V., Breusov, O. N., Dremin, A. N., Pershin, S. V.: Dokl. Vses. Symp. Impul'snym Davleniyam, 1st, 1973, *2,* 50 (1974)
225. Adadurov, G. A., Breusov, O. N., Dremin, A. N., Lazarev, A. I.: Fizika Goreniya i Vzryva *2,* 130 (1966) [English translation: Combustion Explosion and Shock Waves *2,* 82 (1966)]
226. Samsonov, G. V., Alekseevskii, V. P., Bozhko, S. A., Yarosh, V. V.: Poroshkovaja Metallurgija *1,* 93 (1972)
227. Kurdyumov, A. V., Frantsevich, I. N.: Sov. Phys. Dokl. *20,* 235 (1975)
228. Kurdyumov, A. V., Pilyankevich, A. N., Alekseevskii, V. P., Yarosh, V. V.: Sov. Phys. Tech. Phys. *20,* 128 (1975)
229. Bavina, T. V., Breusov, O. N., Dremin, A. N., Pershin, S. V.: Fizika Goreniya i Vzryva *5,* 773 (1975) [English Translation: Combustion Explosion and Shock Waves *5,* 660 (1975)]
230. Gomon, G. O., Rovsha, V. S., Shemanin, V. I.: Dokl. Akad. Nauk SSSR *213*(2), 306 (1973)
231. Deribas, A. A., Nesterenko, V. F., Staver, A. M.: Fiz. Gorenija i Vzryva *8,* 2 (1972)
232. Key, M. H.: Nature *316,* 314 (1985)
233. Cezairliyan, A., Miller, A. P.: Bull. Am. Phys. Soc. *32,* 608 (1987)
234. Van Thiel, M., Ree, F. H.: J. Appl. Phys. *62,* 1761 (1987)
235. Brown, W. L., Freeman, R. R., Raghavachari, K., Schluter, M.: Science *235,* 860 (1987)
236. Heath, J. R., Zhang, Q., O'Brien, S. C., Curl, R. F., Kroto, H. W., Smalley, R. E.: J. Am. Chem. Soc. *109,* 359 (1987)
237. El Goresy, A., Donnay, G.: Science *161,* 363 (1968)
238. Van Vechten, J. A., Keszler, D. A.: Phys. Review B *36,* 4570 (1987)
239. Weisburd, S.: Science News *132,* 101 (1987)
240. Van Thiel, M., Ree, F. H., Grover, R.: Lawrence Livermore National Laboratory Report, UCRL-96492, July 1987 (preprint)
241. Van Thiel, M.: Written communicaton (17 Sept 1987)
242. Akashi, T., Sawaoka, A.: US Patent 4,695,321 (assigned to New Mexico Tech Research Foundation, Socorro, NM)

Surface-Tension-Driven Flow in Crystal Growth Melts

Dietrich Schwabe

I. Physikalisches Institut der Justus-Liebig-Universität, Heinrich-Buff-Ring 16, 6300 Gießen, F.R.G.

Convection is an important parameter in crystal growth from the melt. A driving force for convection besides buoyancy, crystal rotation and crucible rotation, is the surface-tension gradient due to temperature or concentration gradients in the free melt surface. This article points out that in the various melt-growth techniques with free melt surface, e.g. the floating-zone technique, the Czochralski technique and the open-boat zoning, surface-tension-driven flow is likely to occur and to influence the growth. Surface-tension-driven flow is especially important in small melt volumes, near the free surface of larger volumes, or under conditions of weightlessness (e.g. in Spacelab). Under the large temperature gradients encountered in crystal growth, surface-tension-driven flow can become time dependent. Numerical studies, experiments on surface-tension-driven flow, and observations in real growth systems are reviewed.

1 Introduction

The growth of crystals from the melt is one of the most important crystal-growth techniques. Large bulk crystals can be grown at relatively high growth speeds compared to other methods. Therefore the melt-growth technique is applied whenever possible. Successful growth requires the control of heat and mass fluxes from and to the growing crystal interface. These fluxes are carried by diffusion and convection. Reference[1)] gives an introduction to the problems of mass and heat transport in crystal growth. Convection in the melt is always of importance if diffusion is slow, which is the case for mass diffusion. The rôle of convection is more complicated in heat transport where we can have melts which are good thermal conductors such as metals or semiconductors with high thermal diffusivity and, less good thermal conductors like oxides with low thermal diffusivity. Because of the convective part of heat and mass transport, crystal growth requires a clear understanding of, and, if possible, control of convective processes.

In crystal-growth melts we can have forced convection (e.g. rotation of the crystal or of the crucible in the Czochralski technique) and natural convection. Natural convection can be buoyant convection by heating from below or heating from the side or can be Marangoni convection (surface-tension-driven convection). Buoyant convection is driven by density differences in the gravitational field of the earth and is a volume phenomenon. Marangoni convection is driven by surface tension differences in the free liquid surface and is a purely surface effect. Therefore buoyant forces are strong in large volumes, whereas surface-tension-driven convection can dominate in small volumes with free liquid surface. A free melt surface is a decisive advantage in many growth techniques to avoid heterogeneous nucleation or to grow from pure melt e.g. in floating-zone growth, in open-boat zoning, in Czochralski growth, in ribbon growth, in Verneuil growth, and in laser melting. Temperature or concentration gradients are maintained during growth. Therefore the conditions for surface-tension-driven flow are satisfied in many melt growth techniques.

2 The Principle of Surface-Tension-Driven Convection

The principle of surface-tension-driven convection is discussed now for the case of thermal driving forces. The surface tension σ is temperature dependent and decreases with increasing temperature. In a free surface with a temperature gradient an imbalance of

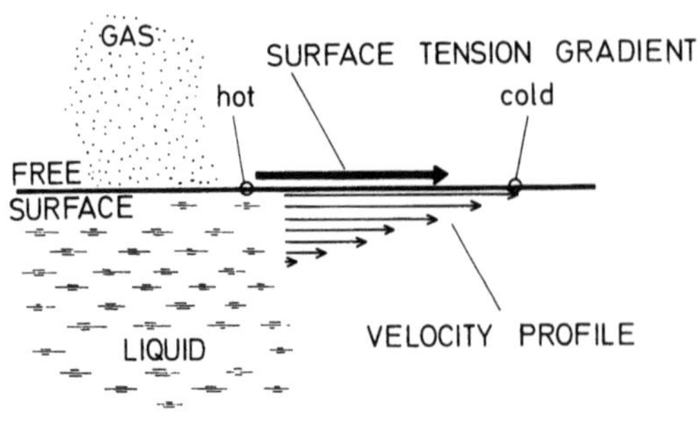

Fig. 1. The mechanism of surface-tension-driven convection

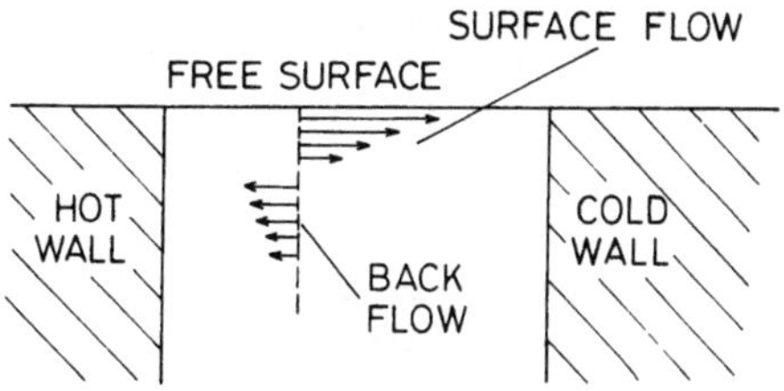

Fig. 2. Thermocapillary convection due to a temperature gradient parallel to the free surface

surface-tension forces exists which drives areas of low surface tension (hot) to areas of high surface tension (cold). The surface motion is balanced by viscous sheer and is transfered to neighboring liquid layers because of the fluid's viscosity (Fig. 1). Because of continuity we have a backflow in the bulk liquid. The convection goes on as long as a temperature gradient exists in the free surface.

Convection driven by temperature or concentration gradients in free liquid surfaces is also known as the Marangoni effect. Marangoni effects are phenomena of general importance for all free liquid surfaces. These effects have been identified as playing a rôle in the convection of fluid layers that have a free surface heated from below or cooled from above, in evaporation, in liquid-liquid extraction, in spreading of films, in boiling heat transfer, in bubble motion, in two-phase flow and in many other fields of small scale hydrodynamics. The most relevant problems and literature have been discussed in[2–8]. It seems that crystal growers only recently became aware of convection driven by surface-tension gradients[9]. This arose because of the possibility of growing crystals under reduced gravity in an orbiting spacecraft, in which case buoyancy induced convection will be diminished and Marangoni effects will be of obvious importance. One can distinguish between two ideal cases of pure thermal surface-tension-driven convection:

- if the temperature gradient is parallel to the free surface the flow is called thermocapillary convection (Fig. 2). Every temperature gradient parallel to the free surface gives rise to motion, however small this temperature gradient might be. The onset of thermocapillary convection is given for Marangoni number Ma = 0
- if the temperature gradient is perpendicular to the free surface, a critical temperature gradient (a critical Marangoni number Ma of the order of 70) must be exceeded for the onset of convection. This is the case of the Bénard-Marangoni instability with hexagonal convection cells[10, 11] (Fig. 3).

In crystal growth from the melt we usually have a mixed case with a temperature gradient inclined to the free surface.

The Marangoni number is given by $Ma = -\frac{\partial \sigma}{\partial T} \cdot \frac{\Delta T \cdot D}{\eta \cdot \chi}$ where $\frac{\partial \sigma}{\partial T}$ is the temperature dependence of the surface tension σ, ΔT the applied temperature difference; D is a

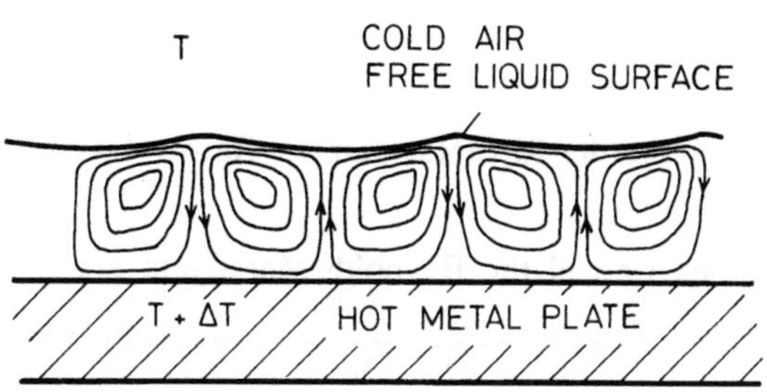

Fig. 3. Marangoni instability due to a temperature gradient perpendicular to the free surface

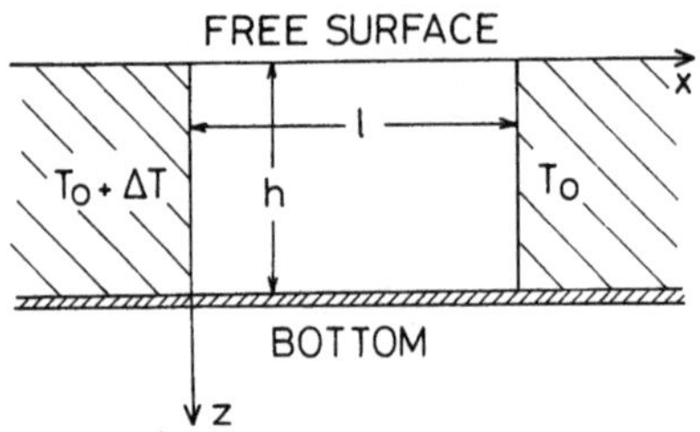

Fig. 4. Geometry and coordinate system

characteristic dimension of the problem, η the dynamic viscosity and χ the thermal diffusivity. The Maragoni number is the most suitable parameter to characterize surface tension driven convection. Ma is determined from a proper non-dimensionalization of the Navier-Stokes differential equation of the hydrodynamic problem or from the boundary conditions at the free surface[12, 13]. This is demonstrated now for a simplified two-dimensional case. Consider a fluid with a free upper surface and with height h between two solid walls a distant l apart, which are at different temperatures (Fig. 4). Whereas the stream velocities v_x and v_z are zero at both side walls and at the bottom we have a tangential stress boundary condition for the surface velocity v_x at the free surface:

$$-\eta \cdot \frac{\partial v_x}{\partial z} = \frac{\partial \sigma}{\partial x} = \frac{\partial \sigma}{\partial T} \cdot \frac{\partial T}{\partial x} .$$

This is the condition which balances the surface tension force against the viscous shear force. We nondimensionalize and normalize all variables x, z and T to get dimensionless units (capital letters),

$$X = \frac{x}{l}, \quad Z = \frac{z}{h}, \quad \theta = \frac{T - To}{\Delta T}, \quad V_x = \frac{v_x}{v_r}$$

where v_r is an unknown reference velocity of the order of v_x. Putting these dimensionless units into the stress boundary condition gives

$$-\eta \cdot \frac{v_r}{h} \cdot \frac{\partial V_x}{\partial Z} = \frac{\partial \sigma}{\partial T} \cdot \frac{\Delta T}{l} \cdot \frac{\partial \theta}{\partial X} .$$

Because $\frac{\partial V_x}{\partial Z}$ and $\frac{\partial \theta}{\partial X}$ are both of the order of unity we get for the reference velocity

$$v_r = -\frac{\partial \sigma}{\partial T} \cdot \frac{\Delta T}{\eta} \cdot \frac{h}{l} .$$

We write the Reynolds number $Re = \frac{v_r \cdot l}{\nu} = -\frac{\partial \sigma}{\partial T} \cdot \frac{\Delta T \cdot h}{\eta \cdot \nu} \equiv M$

which we call the Reynolds-Marangoni number M. The Péclet-Marangoni number is $Ma \equiv M \cdot Pr = -\frac{\partial \sigma}{\partial T} \cdot \frac{\Delta T \cdot h}{\eta \cdot \chi}$ with the Prandtl number $Pr = \frac{\nu}{\chi}$.

There is another valuable dimensionless number to discuss Marangoni effects in earthbound experiments. This is the dynamic Bond number

$$Bo \equiv Gr \cdot M^{-1} = \frac{\varrho \cdot \alpha \cdot g \cdot h^2}{\frac{\partial \sigma}{\partial T}}, \text{ were Gr is the Grashof number, } Gr = \frac{\alpha \cdot g \cdot \Delta T \cdot h^3}{\nu^2},$$

where α is the volume expansion coefficient, g the earth's gravitational acceleration and ν the kinematic viscosity. The dynamic Bond number Bo relates the forces of buoyancy to surface-tension gradients.For $Bo < 1$ surface-tension-driven convection dominates over buoyant convection. Small dynamic Bond numbers can be realized either at reduced gravity or for small liquid heights h. Therefore surface-tension-driven flow becomes important under reduced gravity or in small-scale systems.

It should be stressed that the fluid needs a free and clean surface for surface-tension-driven flow to occur. Surface-tension-driven flow can be hindered by soluble impurities (e.g. surfactants) or by insoluble impurities which can build up a solid surface layer. A solid surface layer from impurities can suppress surface-tension-driven flow totally. For instance water and mercury are known to be very sensitive to surface contamination. They are thus not well suited for experiments on thermocapillary convection.

3 Surface-Tension-Driven Flow in Various Crystal-Growth Techniques

3.1 Surface-Tension-Driven Flow in Floating Zones

An idealized floating zone is a cylindrical column of melt with a free cylindrical surface, held by surface tension forces between two solid rods of the same material. In real crystal growth, the floating zone is never cylindrical for various reasons, namely different diameter of the upper and lower rod, deformation of the free surface by gravity and the heating radio-frequency field and the non-planar solid-liquid interface. However a floating zone has rotational symmetry. Floating-zone silicon growth is described in[14, 15]. The heating in the floating-zone technique is concentrated at the middle equator of the floating zone in order to keep a stable zone length under earth gravity. The method is used to grow high-purity silicon and is of some importance for the growth of high-purity, high-melting-point materials such as W, Mo or Al_2O_3. Convection in the zone is driven by buoyancy forces, surface-tension forces, electromagnetic forces in case of RF induction heating, movement of the liquid zone through the solid, and rotation of the two solid rods. Thus, convection in a floating zone depends very much on the operational parameters and can be very complex.

Chang and Wilcox[9, 16] were the first to point out the importance of thermocapillary flow in a floating zone of silicon. They computed the flow, the isotherms and impurity concentration fields in a non-rotating 1 cm diameter Si floating zone under radiation heating and under electron beam heating. For small values of the driving force, one convection cell was formed in the top half of the zone, and its mirror image in the bottom half (Fig. 5). The thermocapillary flow is not appreciably influenced by buoyancy at earth's gravity or by motion of the zone at the usual rates. It has been shown that

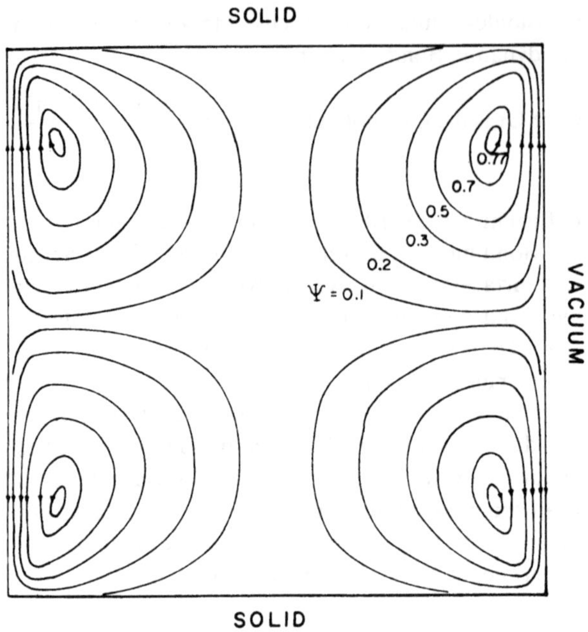

Fig. 5. Computed dimensionless streamlines ψ for surface-tension-driven flow in a floating zone at zero gravity with a parabolic temperature profile on the free liquid surface (after Chang and Wilcox[9])

considerable radial dopant inhomogeneities are introduced with relatively gentle thermocapillary flow[9]. In a later paper[17] it was found that the finite difference equations used in[16] for the fluid adjacent to the free-melt surface satisfy only the boundary conditions and not the Navier-Stokes differential equation. But this did not significantly change the findings of[9, 16].

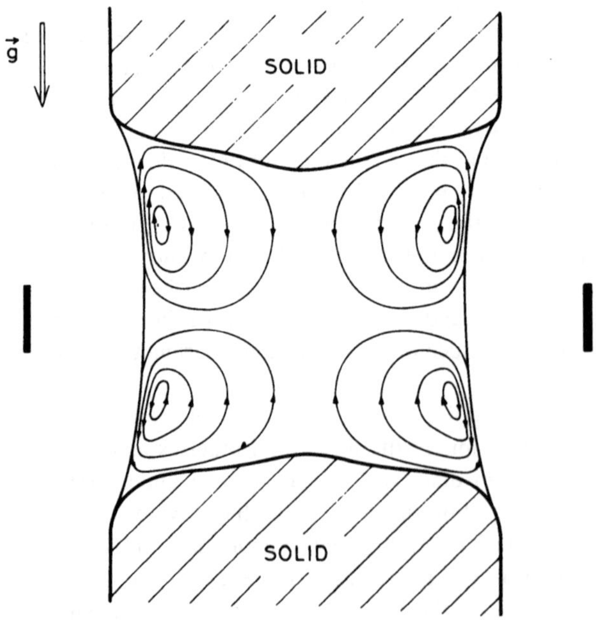

Fig. 6. Streamlines of thermocapillary convection in a radiation heated $NaNO_3$ floating zone (after Schwabe et al.[18])

Figure 6 shows the streamlines of convection in a $NaNO_3$ floating zone[18]. There are many reasons to interpret the streamlines in Fig. 6 as those of flow driven mainly by thermocapillary forces:

1. The picture of streamlines is very similar to that computed by Chang and Wilcox[9, 16].
2. The streamlines are concentrated at the surface. The center of the vortex is near the free surface which is typical for thermocapillary forces. The flow direction in the free surface is from hot (low surface tension) to cold (high surface tension) in both halves of the zone. Especially in the lower half zone the flow direction is contrary to what could be expected from buoyancy.
3. The measured temperature distribution showed that in both zone halves the flow of hot melt is along the surface to the solid interface.
4. The solid-liquid interfaces of both zone halves have a shape which is formed by hot melt flowing along the surface to the interface.
5. The stream velocity in the surface is about 17 mm s^{-1} compared to 11 mm s^{-1} in the middle of the zone in both zone halves.
6. The dynamic Bond number Bo, which relates the forces of buoyancy and surface tension gradient, equals one for a zone length of $l = 4.8$ mm. $Bo < 1$ is a criterion for thermocapillary forces to dominate over buoyancy, and this is roughly fulfilled in the experiment.

In the same paper an oscillatory state of the thermal surface-tension-driven flow and experiments with concentrationally driven surface-tension flow are described[18].

An interesting experimental model for a floating zone has been investigated by many authors[19–49]. It consists of a cylindrical liquid zone suspended between two cylindrical solid rods (metal or graphite) which are held at different temperatures to produce a temperature gradient along the free cylindrical surface. Heated from above, this configuration can serve as a model for the lower part of a real floating zone; heated from below, as a model for the upper part. The model is therefore called half zone, but it should be stressed that there are significant differences in the thermal and geometric boundary conditions between a real floating half zone in crystal growth and the model half zone. Many features of thermocapillary convection have been studied in the model half zone, and these will now be described. As an example we discuss the work of Schwabe and Scharmann[23, 43, 49] and Preisser, Schwabe and Scharmann[36] in more detail.

These authors used sodium nitrate melt as zone liquid. $NaNO_3$ melt is fully transparent and enables the fluid flow to be made visible with tracer particles. The Prandtl number ($Pr = 8.9$) is in the range of some crystal growth melts (oxides) and, due to the "high" working temperature (melting point = 306.8 °C) of the fluid, the thermal boundary conditions of melts are better simulated than with the silicone oils normally used in hydrodynamic experiments. The zones are heated from above to exclude buoyant convection in a vertical, unstable temperature gradient. The steady thermocapillary convection in such zones heated from above occurs in the form of a single axisymmetric toroidal convection roll with hot fluid streaming down in and near the free surface in a thin layer (Fig. 7a). Because of continuity there is an upward back-flow in the bulk volume. Suppressing thermocapillary forces by a viscous oil film causes the flow velocity in the zone to drop considerably (Fig. 7b). The remaining flow is buoyant, driven by the horizontal (radial) temperature gradient. The strong contribution of thermocapillary convection to the heat transport can be seen in Fig. 8; fast surface flow starts from the rim of the upper rod, draws off the thermal boundary layer and bends the isotherms downwards, so that

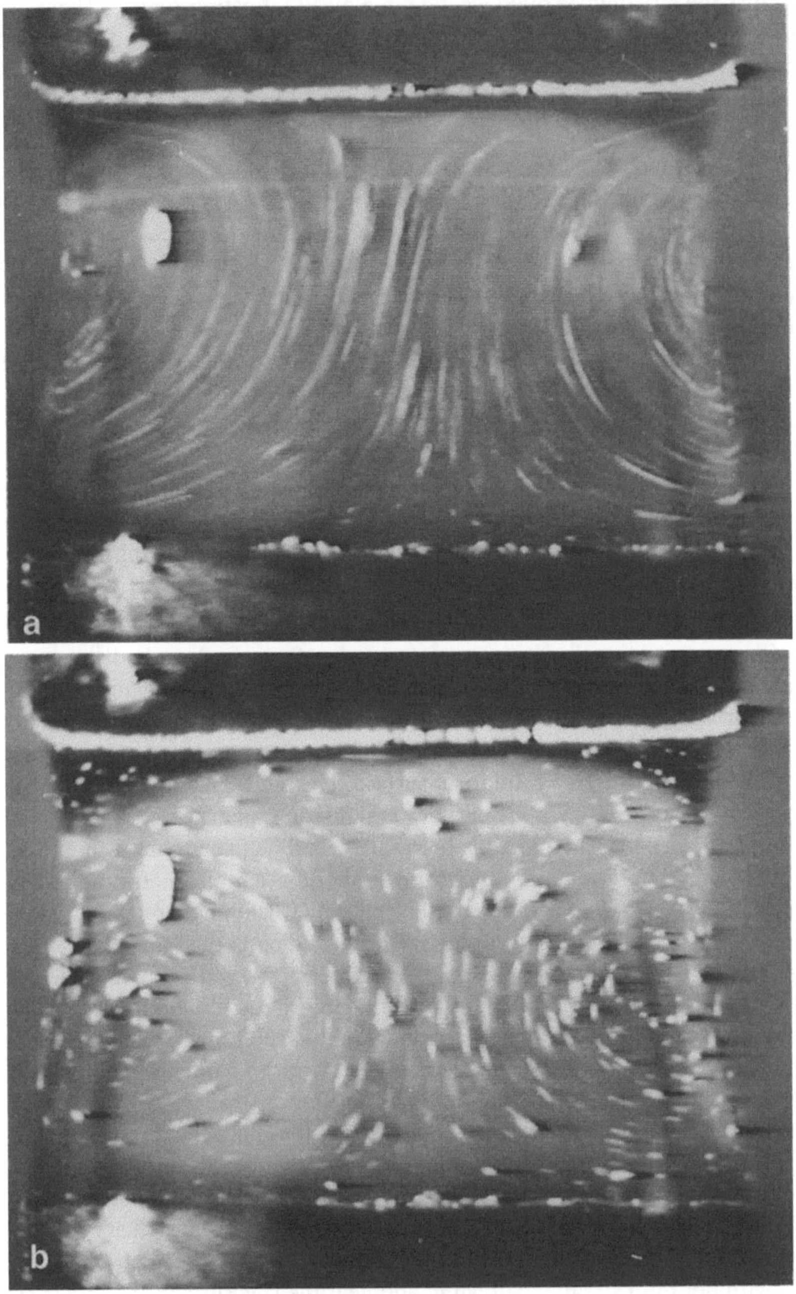

Fig. 7 a, b. Comparison of thermocapillary and buoyancy-driven flow in a $NaNO_3$ zone heated from above (Ø = 6 mm, length = 4 mm, T_{upper} = 360 °C, T_{lower} = 350 °C, both exposure times = 1/8 s). The flow is upwards in the middle in both cases. Compare the streak length of the tracer particles. **(a)** Thermocapillary forces and buoyancy forces in a zone with clean surface (velocity in the free surface about 15 mm s^{-1}). **(b)** Same heating conditions but thermocapillary forces suppressed by an oil film of approximately 50 μm thickness (maximum velocity about 3 mm s^{-1}), (after Preisser et al.[36])

the surface is hotter than the interior of the zone. Hot fluid flows downwards against the buoyant forces, which indicates that the observed flow is thermocapillary.

In Fig. 9 the radial velocity distribution of steady thermocapillary flow is given at axial position Z = 0.68. It can be seen that the downward surface stream is restricted to a narrow boundary layer with a maximum flow speed of 11 mm s^{-1}. Furthermore the back-flow is limited to a region near the free surface with a maximum back-flow speed at R = 0.35. This becomes more obvious in zones with larger diameter[36)]. Such a localization of thermocapillary flow near the free surface would give rise to radial inhomogeneities of dopants in floating zone crystal growth.

It was found in glass melts[48)], in silicon oils[40)] and in $NaNO_3$[36)], that the stream velocity of thermocapillary convection increases linearily with increasing temperature difference. This dependence is different from that of buoyancy-driven flow where the stream velocity rises with the square root of the temperature difference.

The steady state of thermocapillary convection has been studied in floating half zones of silicone oils in a microgravity environment in a sounding rocket experiment[41)] and in a Spacelab experiment[45, 47)]. Under microgravity conditions, pure thermocapillary convection can be observed. The agreement between calculations and the experiments was quite satisfactory[41)]. More calculations of thermocapillary flow in floating zones can be found in references[50–66)]. Among these, reference[50)] in particular, treats the half-zone configuration and high Marangoni numbers. The flow pattern and isotherms calculated[50)] correspond to those measured[36)].

In the floating half zones heated from above, thermocapillary convection becomes time dependent (oscillatory) for larger Marangoni numbers. The transition from steady to oscillatory flow occurs at a critical Marangoni number Ma $\approx 10^4$. This has been observed in half zones of $NaNO_3$[18, 23, 36)], alcohol[67, 68)], KCl[24)], octadecane[20)], Fluorinert[27, 37)] and hexadecane[37)]. Time-dependent flow would give rise to dopant inhomogeneities (striations) in crystal growth and it might be important to avoid this oscillatory state. In the oscillatory state the otherwise axisymmetrical flow field is of lower symmetry and the perturbation of streamlines travels around the zone axis as a

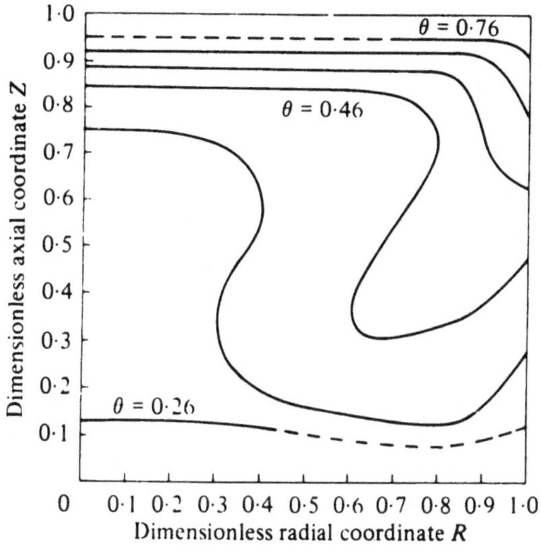

Fig. 8. Isotherms in the vertical right half of the zone shown in Fig. 7a (conditions as in Fig. 7a), (after Preisser et al.[36)])

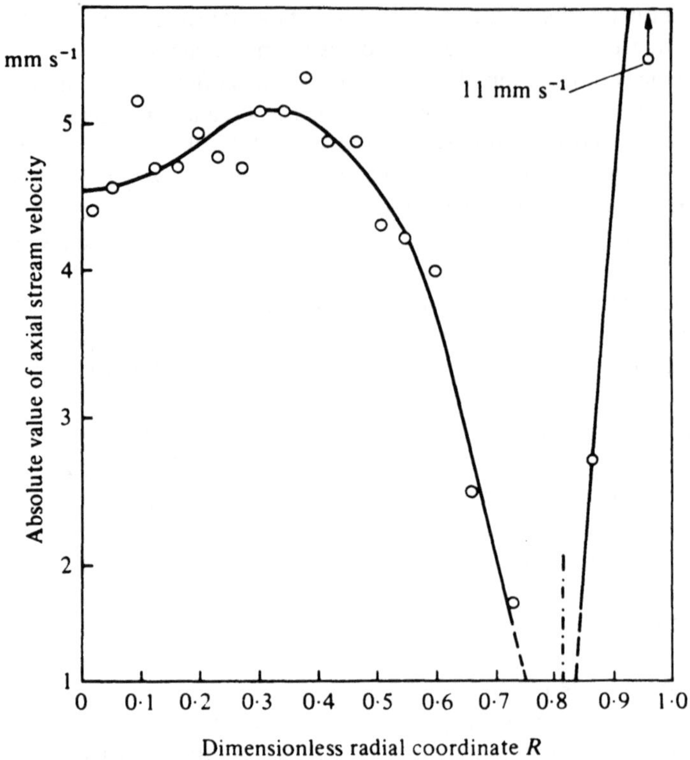

Fig. 9. Radial velocity distribution of thermocapillary flow at zone height Z = 0.68 (l = 4.1 mm, radius = 3.0 mm, $\Delta T = 10$ K, Ma = $4.8 \cdot 10^3$, the *dashed line* indicates the position of the roll axis), (after Preisser et al.[36])

running wave[36]. This has been made visible in a central vertical light cut and has been measured with the aid of thermocouples. Figure 10 shows the oscillations of the streamlines recorded by the tracks of illuminated tracer particles in the light cut. In Fig. 10 the instant is depicted when the vortices in the light cut have extreme positions. The vortex in one half of the zone section appears smaller than the opposite vortex, and the centre of the smaller vortex is located nearer to the lower solid boundary. The larger vortex is characterized by a centre closer to the upper boundary. In the light cut the time dependence is observed as periodical interchange of the shape of the vortices in the left and right parts of the zone. After one half of the oscillation period, the left vortex and the right vortex have changed their positions. While the features of the steady thermocapillary convection roll are independent of the azimuthal angle, the oscillatory state at a fixed time is characterized by a sinusoidal distortion in azimuthal direction of both the roll cross-section and the roll axis. The time dependence is caused by the propagation of this distortion around the zone axis. Therefore the instability is recorded as a flow oscillation in the fixed light cut or as a temperature oscillation with a fixed thermocouple. The rotational symmetry of the zone selects an integer number n of periodical roll distortions with wavelength λ on the circumference of the zone: $n \cdot \lambda = 2\pi a$, where a is the zone radius. The above description corresponds to the mode n = 1 which is shown in Fig. 10.

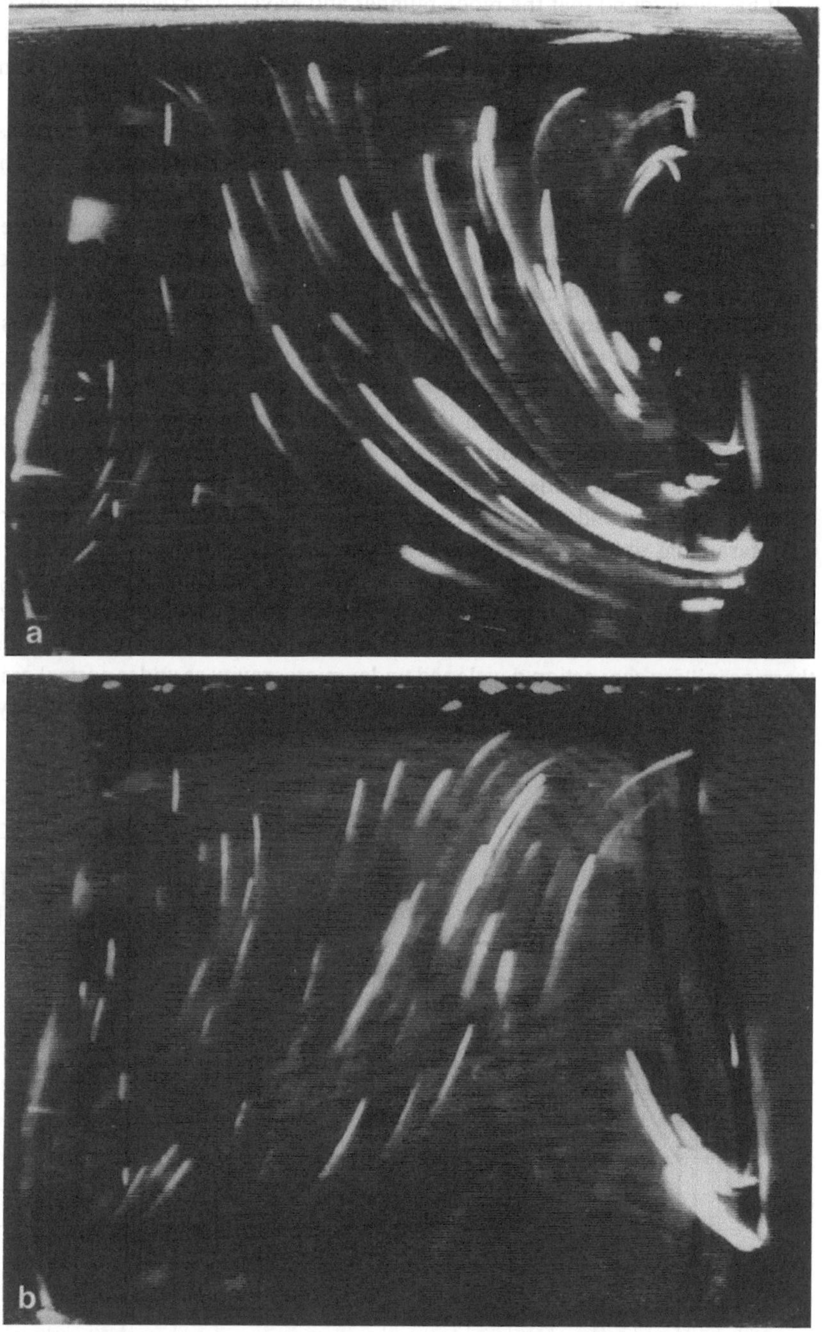

Fig. 10 a, b. Oscillatory thermocapillary convection in a 6 mm diameter $NaNO_3$ zone heated from above. The oscillation period is about 2 s, the pictures are taken with time interval of approximately 1 s (after Preisser et al.[36])

It has been found that the mode number and wavelength depend on the aspect ratio, A = l/a, of the zone. The mode number increases with decreasing aspect ratio. Higher modes n of oscillatory flow have been observed and the frequency f, mode number n, and critical Marangoni number Ma^c of the transition from steady to oscillatory flow have been measured[36]. A phase shift between flow and temperature oscillations has also been measured[22], and the origin of the oscillatory state has been discussed[67, 37]. The mechanism of oscillatory thermocapillary flow has not yet been clarified.

The instabilities of dynamic thermocapillary liquid layers have been studied theoretically[69, 70] and have been applied to floating zones[71]. Propagating, so called hydrothermal waves have been found as instabilities but the calculated critical Marangoni number is one order of magnitude smaller than the measured one. End effects present in the experiment but not in the calculations are discussed as a possible reason for the discrepancy and when this is taken into account, calculated critical Marangoni numbers come closer to the measured ones. But than the calculations predict the wrong frequency.

The gravity-independent driving force of thermocapillary convection allows the oscillatory state to occur under microgravity conditions. The thermocapillary convection became oscillatory under microgravity in a sounding rocket experiment for two supercritical Marangoni numbers[30]. Figure 11 shows the temperature oscillations of oscillatory thermocapillary flow as measured in a $NaNO_3$ half zone under microgravity. The frequency spectrum of the temperature oscillations under low gravity was very similar to that under earth gravity. From this it is concluded that the same oscillatory flow state of thermocapillary convection developed under microgravity as under normal gravity. Thermocapillary convection in a half zone needs no interaction with buoyancy forces to

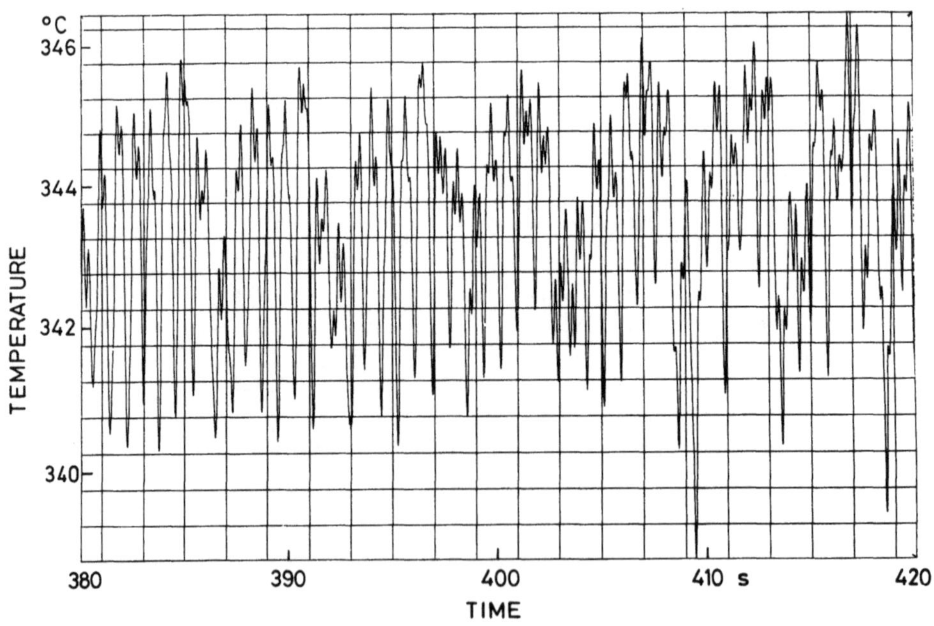

Fig. 11. Temperature oscillations under 10^{-4} g in a $NaNO_3$ floating half zone heated from above with $Ma = 3.6 \cdot 10^4$ (after Schwabe et al.[30])

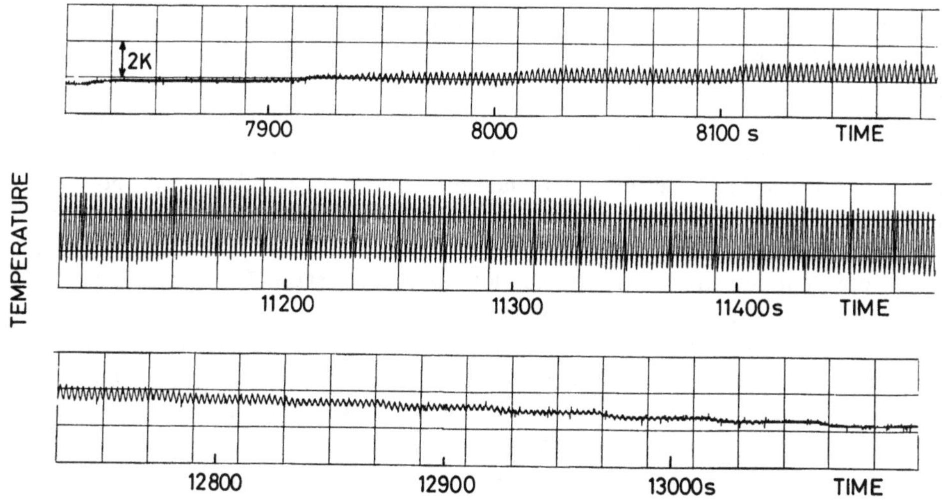

Fig. 12. Measurement of the critical Marangoni number for the onset of temperature oscillations under microgravity (10^{-4}g) in a $NaNO_3$ floating half zone heated from above. *Top:* onset of oscillations during increase of Ma (Heating up), *middle:* fully developed oscillations at high supercritical Ma, *bottom:* fading of oscillations during decrease of Ma (cooling down) (after Schwabe and Scharmann[43])

become instable. Indications that the flow becomes turbulent at higher Marangoni numbers have been found in a further microgravity experiment in a sounding rocket[42].

The critical Marangoni number Ma^c for the transition from steady to oscillatory thermocapillary convection in floating half zones has been measured in a sounding rocket experiment[49] and in a Spacelab experiment on board a free-flying satellite[43]. In the sounding rocket experiment[49] the critical Marangoni number ($Ma^c = 8500$) was smaller that that measured in the reference experiment under normal earth gravity ($Ma^c = 10\,200$). The frequencies of the temperature oscillations are smaller under microgravity than under normal earth gravity. This gravity dependence may be due to the stabilizing effect of heating from above in the laboratory experiments.

In the experiment on board a satellite, the critical Marangoni numbers under normal and microgravity are of the same order of magnitude with a tendency to smaller Ma^c under microgravity for longer zones[43]. Figure 12 shows temperature oscillations under microgravity near the transition point from steady to oscillatory flow. Normally the oscillation frequency increases with increasing Marangoni number, but in this experiment the opposite has also been observed. Frequency jumps and beat frequencies have also been found in the microgravity experiment and under normal gravity.

In crystal growth from the melt using the floating zone technique, rotation of the rods is normally applied in order to achieve an axisymmetric temperature distribution and liquid mixing of the zone. Rotation induces forced convection in the floating zone which interacts with thermocapillary convection. This has been experimentally studied by Chun[21,28]. Counter-rotation and iso-rotation of the two solid end rods have been found to reduce the thermocapillary convection. A suppression of oscillatory thermocapillary convection can be achieved by single- or iso-rotation. The effect of fluid rotation on thermocapillary convection was studied numerically for various combinations of the crystal and feed rod rotation: iso-rotation, single rotation, and counter rotation[57]. In the presence of a high fluid rotation, thermocapillary convection is confined to a thin layer near the free surface, whereas forced convection due to rotation occupies the zone interior.

Other authors have computed the interaction of thermocapillary and rotational flow as well[51]. They found a confinement of thermocapillary flow to the surface by uniform rotation. For realistic values for silicon, however, the thermocapillary flow is not linear: it is strong and unsteady and penetrates the zone interior. Since silicon has a large value of thermal conductivity, the temperature fields are expected to be determined by conduction alone, but for stronger nonlinear flow the results show that convection is also important for the temperature distribution[51].

Modern silicon floating zones are strongly necked-in and are induction heated by a radio frequency coil. The convection pattern in such real silicon floating zones induced by the electrodynamic forces has been studied numerically[72]. The authors show by non-dimensional analysis that electrodynamic convection is orders of magnitude stronger than surface-tension-driven flow. This is confirmed by experimental work[73]: The radial Si distribution in Fe-3 wt% Si crystals grown by floating zone melting with induction heating differs remarkably from that with electron beam heating. From the radial Si distribution the main features of the flow pattern in the molten zone for both heating methods have been deduced. Surface-tension-driven flow was shown to be dominant in the case of electron-beam heating. With induction heating the flow direction was found to be inverted in comparison with electron beam heating. The change of the flow direction is explained by the action of electrodynamic forces which drive the melt convection in the opposite direction from that of thermocapillary forces.

The significance of surface-tension-driven flow is still a subject of controversy amongst the crystal-growth community. Numerical calculations and the model experiments on the one hand, clearly show its importance; on the other hand the temperature coefficients of surface tension (e.g. $\sigma = -4.0 \cdot 10^{-4}\ \mathrm{N \cdot m^{-1} \cdot K^{-1}}$ for Si) might be lower than anticipated, and a dirt skin at the free surface could suppress thermocapillary forces. However, silicon growth for instance is carried out under vacuum or argon; SiO_2 is fairly soluble in Si, and the dynamics of surface motion and fluid motion make a closed skin improbable. However, some of the recent crystal-growth publications do consider the effects of surface-tension-driven flow.

In floating-zone growth of 4 mm tungsten and molybdenum crystals with electron-beam heating under high vacuum without crystal rotation, observations have been made which are easily understood by taking surface-tension-driven flow into account[74]. Radial inhomogeneities are observed for zone length $\leq$ 3.5 mm with distribution coefficient nearer unity in the middle of the crystal, because for this aspect ratio, thermocapillary convection and strong mixing are confined to a surface region. For zone lengths exceeding $\approx$ 3.5 mm, these radial inhomogeneities vanish, but striations (fine, more or less periodic inhomogeneities in growth direction) appear, because the critical Marangoni number Ma^c is exceeded. The Ma^c for the transition from steady to oscillatory flow and the frequency of the oscillations are of the expected order of magnitude. The phase shift of opposite striations is near π, which indicates mode $n = 1$, consistent with the aspect ratio between 0.5 and 1. In a further paper on tungsten-doped molybdenum single crystals[75] grown by electron-beam floating-zone melting, impurity distributions and calculations are described which show the existence of internal eddies as well as surface eddies. These internal eddies, although of minor strength and not observed in the half-zone model experiments with this aspect ration, appreciably affect the impurity distribution in the grown crystal. But one should not forget that the half-zone model experiments are carried out with high Prandtl number fluids whereas molybdenum melt is a low Prandtl number fluid. Therefore the experiments are not fully comparable with respect to the number and size of eddies. Better comparison of model experiments and real growth is possible for experiments with $NaNO_3$[18)] and growth of YIG[76)] because of comparable heating conditions and Prandtl numbers. The radial composition profile and the interface curvature of the YIG crystals grown in a mirror furnace, indicated that thermocapillary convection dominated in the peripheral region[76)].

Silicon was grown under destabilizing temperature gradients (heating from below) in containers without any free melt surface[77)]. Striations vanished in Si when the Rayleigh number Ra was lower than the critical value predicted by theory. But crystallization of Si from melts with a free surface (recrystallization of a hanging drop) gave striated crystals even for Ra $\approx$ 0, thus indicating oscillatory thermocapillary convection even for small Marangoni numbers. In this experiment, temperature gradients parallel and perpendicular to the free surface existed, thus making possible thermocapillary convection and Bénard patterns at the same time. Kölker has recrystallized a hanging drop of silicon under reduced gravity in a mirror furnace during the Spacelab 1 mission[78)]. The drop was rotated at 10 revolutions per minute and was withdrawn from the hot region at a rate of 1 mm/min. The sliced and etched space-grown crystal revealed fine nonrotational striations separated by about 10 μm from each other, as well as strong rotational striations. The fine nonrotational striations are due to time-dependent surface-tension-driven flow. Another interpretation of these fine striations as transients of the rotationals has been excluded by repeating this experiment without rotation during the Spacelab D1 mission: without rotation the same pattern of fine striations appeared[79)].

There is a series of experiments with silicon floating zones under microgravity which show the significance of oscillatory thermocapillary convection in producing dopant striations[80–83)]. A phosphorus-doped silicon crystal of 10 mm diameter has been grown by the floating zone method in a mirror furnace at a rate of 5 mm/min aboard Spacelab 1[80)]. The patterns of nonrotational striations of the space-grown and of the terrestrial reference sample are very similar. Thus, time-dependent thermocapillary flows, driven by gravity-independent surface-tension gradients in the molten zone, appear to be the dominant cause of these dopant inhomogeneities in space. The similarity between the microstriation pattern of the space-grown and of the terrestrial reference crystal suggests that the cause of microstriations on earth is also time-dependent thermocapillary flow. The fact that the distances between the striations are not periodic, but irregular, is probably due to the existence of turbulent thermocapillary flows under the extreme conditions in silicon melts.

A phosphorus-doped silicon crystal of 8 mm diameter has been grown by the floating-zone method in a mirror furnace at a rate of 5 mm/min during the weightless phase of a rocket flight[81)]. From the occurrence of similar microstriation patterns in the microgravity crystal and the terrestrial reference crystal it is concluded that the dopant inhomogeneities in floating-zone silicon crystals – at least for the geometry and crystal size employed here – are caused by oscillatory and turbulent modes of thermocapillary flow in the melt. These two experiments on floating-zone crystal growth of silicon under microgravity show the existence of nonsteady surface-tension-driven flow under microgravity in real growth systems, which has also been demonstrated by model experiments with floating half zones[30, 42, 43, 49]. Moreover, they suggest that the cause of striations in earth-grown crystals is also nonsteady surface-tension-driven flow. To verify this, Eyer and Leiste have carried out crystal growth from floating zones with coated melts, i.e. from melts without free surface[82]. The expectation is that a coating of the melt would suppress surface-tension-driven flow and any nonsteadiness of it. Indeed, silicon crystals with 10 mm diameter grown by the floating-zone technique in a double ellipsoid mirror furnace with SiO_2-coated melts, have been found to be free of striations. This proves that thermocapillary convection is the cause of the microstriations which one always finds in uncoated crystals of this size.

The same authors report in their paper on the stability of SiO_2 skins on silicon melt[82]. They produced an approximately 5 μm thick SiO_2 layer on the single-crystalline feed rod by thermal oxidation at 1100 °C for 50 h in flowing oxygen (about 1 atm) saturated with water.Such an amorphous SiO_2 layer formed a continous elastic and transparent skin of the molten zone. Under the thermal conditions of zone melting, chemical reaction of the skin is sufficiently slow to keep the skin complete for about 6 min. Increasing the contact time causes holes to form in the skin. The holes increased slowly with time, presumably by the formation of SiO which evaporates. Reaction under the skin which would lead to high levels of dissolved oxygen in the growing crystal cannot be excluded. These observations show that an oxide film is not stable for longer reaction times on silicon melts. The silicon melts in floating-zone growth, in Czochralski growth and in ribbon growth should thus be free of closed oxide skins and thermocapillary forces should be active in driving the flow.

A closed oxide skin on the Si melt is an effective means to suppress thermocapillary forces[83]. This was shown by recrystallizing a phosphorus-doped silicon crystal under microgravity by float-zoning aboard a sounding rocket at a rate of 5 mm/min. The crystal was coated with SiO_2 to suppress thermocapillary forces. It grew free of microstriations. During its travel, the molten zone was counterdoped with boron released from a point-shaped depot. Analysis of the boron distribution in the solidified melt showed that under microgravity the dopant propagation is purely diffusive. The SiO_2 skin suppressed thermocapillary flow totally and therefore there was no convection at all under microgravity.

A case of time-dependent surface-tension-driven flow many have been observed in space grown InSb[84]. InSb has been melted and recrystallized in quartz ampoules, but over a length of approximately 20 mm the melt diameter was smaller than the ampoule diameter and the crystal grew into a "floating zone" (melt with free surface). In the region with free surface, surface striations of approximately 1 Hz, which correlate with dopant striations, have been found. These striations are not found when growth proceeded with the melt having no free surface, which is strong evidence that the observed time dependence of growth was not caused by external perturbations. In a similar experiment with InSb: Te, the same "floating zone" occurred in the quartz tube and the measured Te distribution in the grown crystal has been calculated satisfatorily by taking into account convective mixing (thermocapillary convection) in the region with free surface[85].

3.2 Surface-Tension-Driven Flow in Czochralski Growth

Crystal growth from the crucible by the Czochralski technique ist the most widely used method for research and industrial production. The method, literature and hydrody-

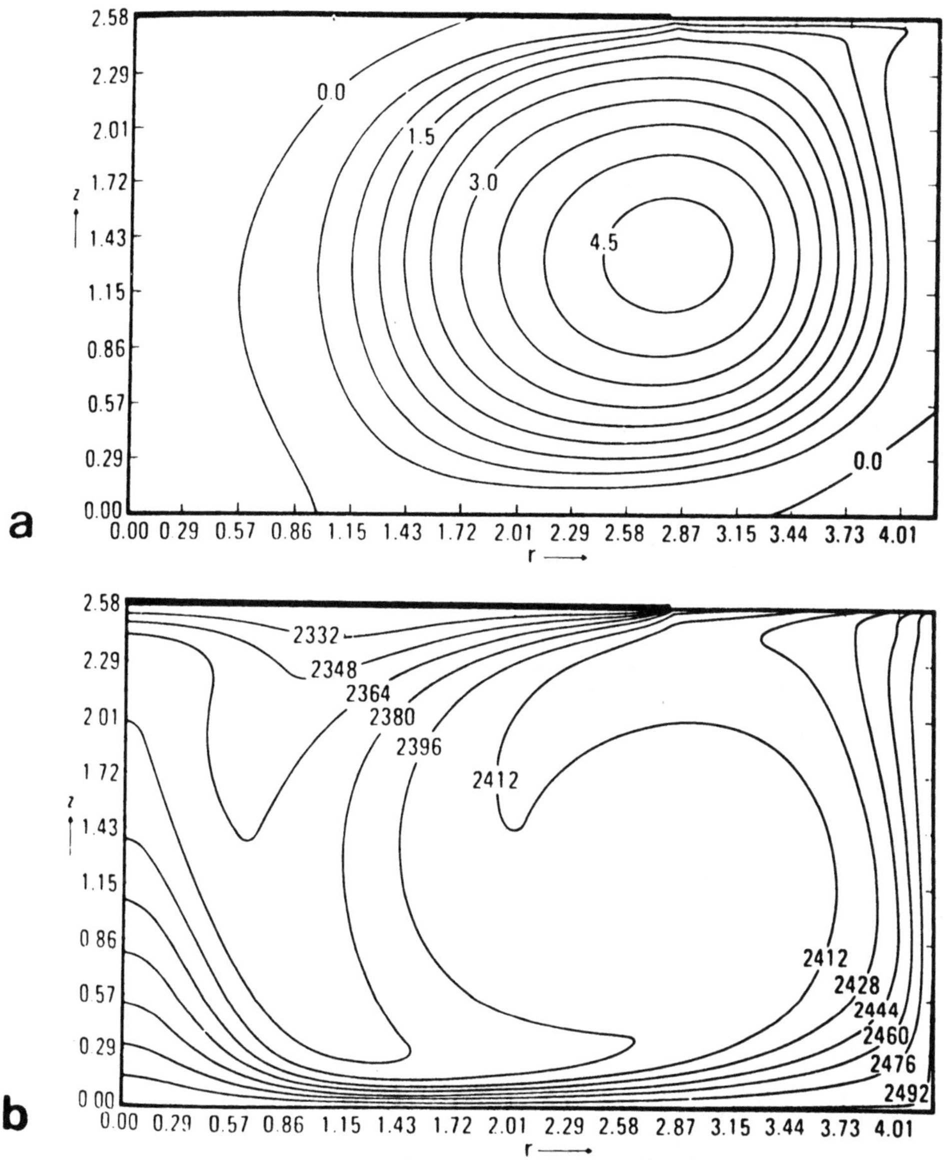

Fig. 13 a, b. The flow field of thermocapillary convection: (**a**) streamlines, (**b**) isotherms in Al_2O_3 melt under microgravity (10^{-4} times normal earth gravity) in the right part of the crucible (after Langlois[91])

namic problems have been discussed in recent review articles[86–90] where more literature can be found. The basic principle of the Czochralski method is to melt the material to be crystallized in a crucible and to extract the growing crystal from the free melt surface. The heating and cooling conditions of a Czochralski crucible generate a radial temperature gradient within the melt and along the free surface thus giving rise to buoyancy forces and thermocapillary forces. Besides the radial temperature gradient there is an axial temperature gradient because the melt is cooled from above. Rotation of the crystal

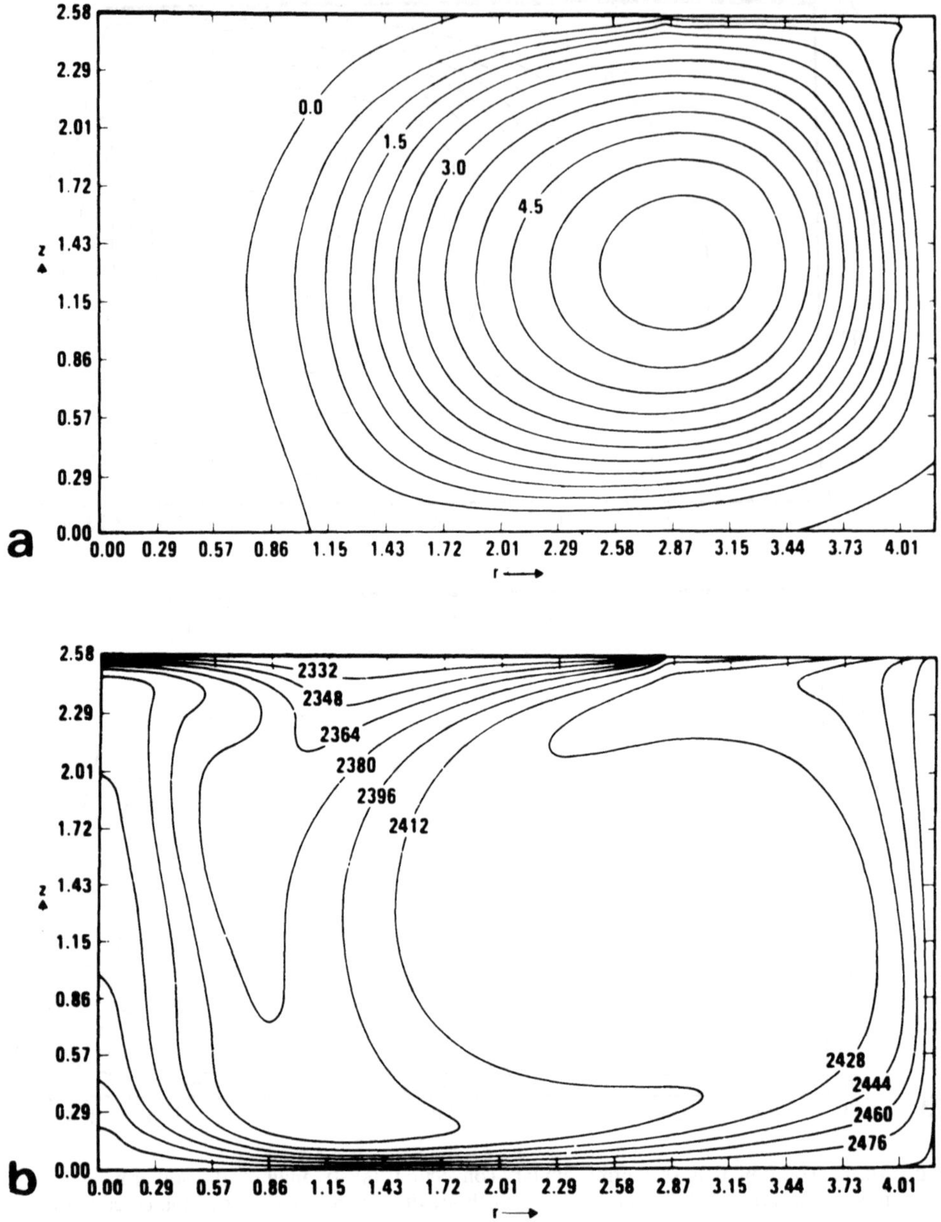

Fig. 14a, b. The flow field in the right part of the crucible of combined buoyant and thermocapillary convection: **(a)** streamlines, **(b)** isotherms (after Langlois[91)])

and of the crucible can further complicate the flow in the crucible. Because of the smaller free-surface-to-volume ratio in Czochralski growth compared to floating-zone growth, surface-tension-driven flow is of minor importance in a crucible. But the crystal is extracted from the melt surface which is the region where surface tension gradients act, thus gaining importance in the near flow field of the crystal-melt interface.

Langlois[91] was the first to pay attention to thermocapillary forces in numerical studies of Czochralski systems. He computed the flow of Al_2O_3 melt in a crucible with 4.191 cm radius and 2.580 cm melt depth from which a crystal of 2.794 cm radius is grown. Figure 13 shows the streamlines and isotherms in such a crucible under microgravity (10^{-4} g) with crystal temperature of 2316 K and crucible temperature of 2493 K. It is noted that the meridional circulation is quite intense and that the temperature field is convection dominated. Calculations with 10^{-3} g showed that buoyant convection in microgravity is completely swamped by thermocapillary flow. For comparison, the calculation was repeated for normal earth gravity, i.e. with the full buoyancy force. The results are shown in Fig. 14. The strength of the meridional circulation, as measured by the maximum value of the stream function, is only 23% greater than that obtained in microgravity. This does not imply that terrestrial buoyant convection "only adds 23%" to thermocapillary flow, because the effects do not combine linearly. The calculation at normal earth gravity was also done with thermocapillarity turned off. This led to a flow pattern quite similar to that of Fig. 14, but with a meridional circulation 40% weaker. The appropriate conclusion seems to be that, for terrestrial growth of Al_2O_3 of these dimensions and with neither crystal nor crucible rotation, thermocapillary flow is significant but not dominant.

Evidence for the significance of thermocapillary forces in such an oxide melt has been found in experiments on $NaNO_3$ in an iridium crucible[92]. The crucible diameter was 35 mm and the melt heigth was typically 32.5 mm. The authors observed and measured under two conditions of the melt surface: with free and clean surface allowing thermocapillary forces to act, and with a rigid film on the melt surface producing a non-slip condition which suppresses thermocapillary forces. Figure 15 shows the velocity profile of the inward radial flow with a free and clean surface and with a covered surface. A

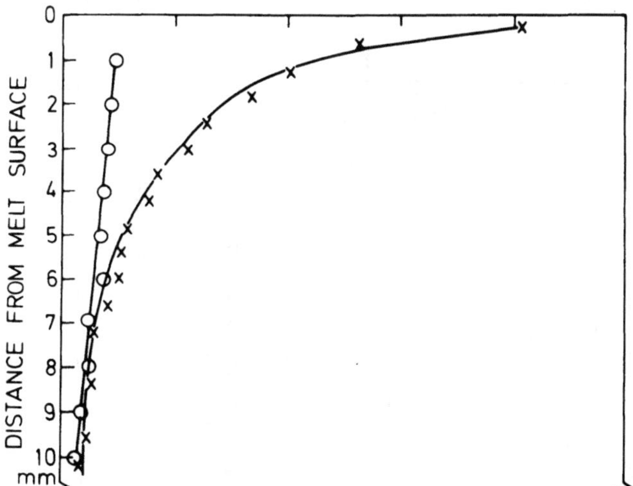

Fig. 15. Comparison of the velocity profiles of inward radial flow for free and clean surface (x) and the surface with rigid film (o). For both cases: $T_{wall} = T_{bottom} = 340\,°C$, melt height = 32.5 mm, "crystal" of quartz glass with 5 mm radius held at 307° C (after Lamprecht et al.[92])

drastic decrease of the maximum flow speed near the surface from 20 mm s^{-1} for a clean surface to 3 mm s^{-1} for a covered surface was found. The velocity profile under the action of thermocapillary forces has boundary-layer character whereas it is approximately linear under the non-slip condition. The boundary-layer character (in this case the steep increase of the radial flow velocity when approaching the free surface) is evidence that the thermocapillary forces act only in the free surface (tangentially inwards). Figure 16 compares the observed stream patterns for the two surface conditions. With thermocapillary forces acting, the streamlines follow the shape of the free surface in both meniscus regions. The flow velocities are higher, and the center of the convection roll is nearer to the free surface compared to the case of non-slip. Temperature distributions for the two surface conditions are compared in Fig. 17. With thermocapillary forces acting, the rapid surface flow transports hot melt from the crucible wall towards the crystal. The 320 °C isotherm bulges towards the crystal. The isotherms are compressed near the crystal meniscus, thus helping to control the crystal diameter. The crystal interface is then expected to be convex towards the melt. With then non-slip condition the overall melt temperature is lower, especially in the surface region (there are heat losses due to radiation and due to convective transport to the ambient atmosphere). In this case the isotherms are fairly equidistant in the surface, and the crystal interface is expected to be concave towards the melt.

Such a non-slip condition for fluid flow has been observed at the surface of freshly molten $LiYF_4$ in an iridium crucible under Hf atmosphere[92]. A layer of oxifluorides and carbon particles formed a solid skin on the surface. Under this condition only slow flow was observed in the bulk melt. When this layer was removed much faster flow developed. Crystal growth from the melt with covered surface was impossible, whereas it was easy

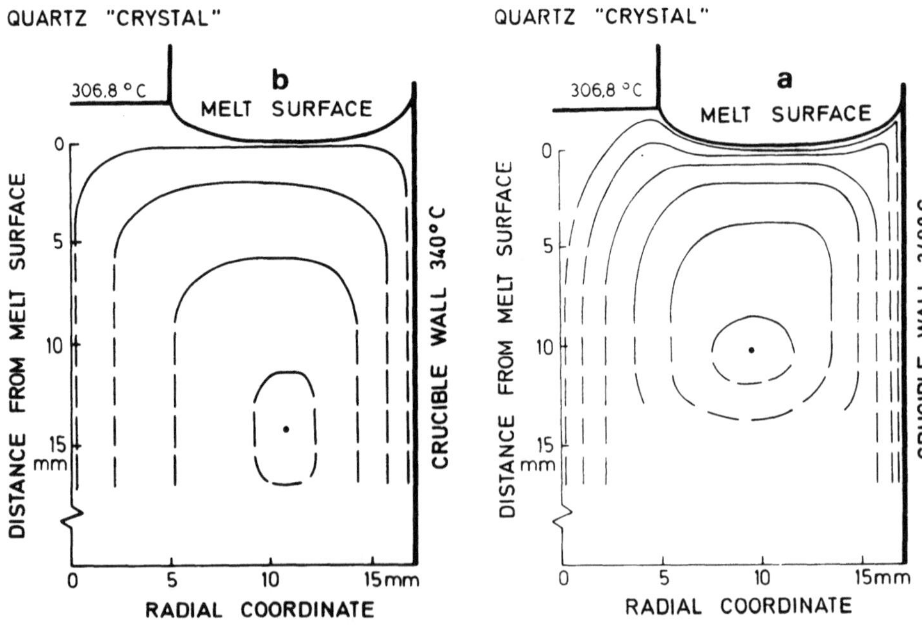

Fig. 16 a, b. Sketch of streamlines for free and clean surface **(a)** and for the non-slip condition **(b)** under the same conditions as in Fig. 15 (after Lamprecht et al.[92])

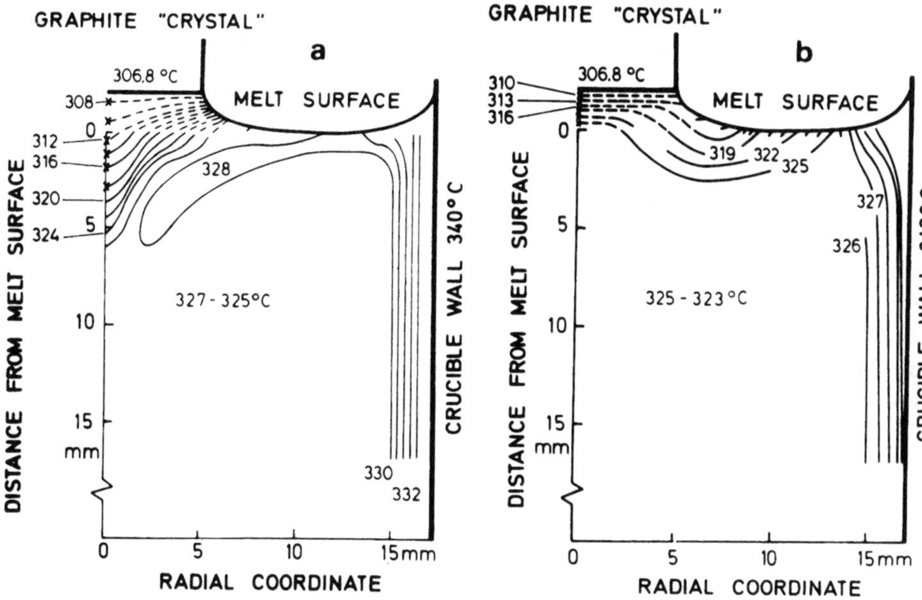

Fig. 17 a, b. Isotherms (°C) in the $NaNO_3$ melt with thermocapillary forces acting in the free and clean surface **(a)**, and with a non-slip condition at the surface **(b)**. $T_{wall} = T_{bottom} = 340\,°C$, melt height = 33 mm, "graphite crystal" with 5 mm radius held at 306.8 °C. The isotherms *(full lines)* are drawn according to a net of data points (0.5 mm × 0.5 mm mesh) measured with a thermocouple. The *broken lines* are estimated. The temperature in the bulk was too homogeneous to draw isotherms (after Lamprecht et al.[92])

from the melt with clean surface. Fluorination of the melt is also faster with a clean surface since the flow in the free surface increases the mass transport between the liquid and the gas.

The calculations and experiments reviewed above have demonstrated contribution of thermocapillary forces to the flow and heat transport in oxide melts, especially near the melt suface and in the region of the crystal meniscus. It is clear that thermocapillary forces influence the interface shape of the crystal. Interface shape transitions of oxide crystals are known to occur when the flow field near the crystal changes from radially inwards (free convection) to radially outwards (forced convection by crystal roataion), caused either by a change of the rotational speed or by a change of the crystal diameter[93–96]. It has been pointed out[92] that thermocapillary forces must be taken into account when considering the problem of fluid-flow transitions leading to interface transitons. Because forced convection by crystal rotation is itself a "surface phenomenon" in the sense that it is driven by shear forces between the rotating interface and the melt, thermocapillary forces, being surface phenomena as well, seem to be of prime importance in the balance of flows near the crystal meniscus.

In the free melt surface in a crucible, temperature and concentration gradients parallel and/or perpendicular to the free surface are possible due to heating and cooling conditions, evaporation, chemical reaction with the atmosphere or the crucible, and impurity enrichment concentration change) by the growth process. Measurements in growth systems indicate radial temperature gradients along the free surface that are one order of magnitude higher than radial temperature gradients in the bulk fluid[97]. Therefore thermocapillary convection is important even for melt heights of several centimeters.

In high-melting-point materials (oxides) grown from noble-metal crucibles, heat flow into the melt through the highly conducting crucible is balanced mainly by emission from the free surface. The melt is strongly cooled from above with the highest vertical temperature gradient in the vicinity (5 mm) of the free surface. Bénard convection cells near the surface, driven by combined buoyancy and thermocapillary forces, that are partly separated from the bulk flow as observed in evaporative convection from the surface of deep vessels, should be the consequence. These surface cells will be distorted by the radial inward flow and will drift towards the crucible center. The pronounced temperature fluctuations observed in the surface region of the melt[99–101] and some features of the spokes patterns[102] should thus be partly due to surface-tension-driven flow effects.

A fluid with a free upper surface heated from below or cooled from above makes a transition from "no flow" to convection rolls or convection cells, when either a critical Rayleigh number Ra^c or a critical Marangoni number Ma^c is exceeded[103]. Figure 18 shows that a combination of both buoyancy and surface-tension-driven forces, can lead to the instability. Pure surface-tension-driven cellular convection (Bénard cells) has been observed on board a spacecraft[104, 105], whereas pure buoyancy driven instabilities can be observed only with a rigid upper boundary. In contrast to this, Jones[106] proposes that the spokes pattern observed on the free surface of some oxide and fluoride melts is caused by a buoyancy-driven instability only. This cannot be the case because the spokes are known to be cooler lines of melt and thus tangential temperature gradients are present in the free surface which induce surface-tension gradients. Because the unstable layer on top of oxide melts has a depth of only a few mm, the instability is of coupled "buoyancy-surface tension" type. As estimated from Fig. 17a[92], one has an unstable fluid layer near the crystal in which we have $Ra/Ra^c = 0.2$ to 8.2 and $Ma^c/Ma = 10$ to 100, showing a dominant contribution from surface-tension forces to the instability in this case.

Miller and Pernell[107–109] have studied fluid flow patterns in simulated garnet melts for which they used water. They noted that in their experiment the liquid height dropped because of evaporation at about 9 mm/h and that the surface cooling effect due to evaporation may replicate in a qualitative way the loss of heat from the surface of a molten garnet by radiation. They found a thin surface layer which was unstable with respect to the thermal gradient (cooled from above) and which exhibited the spokes

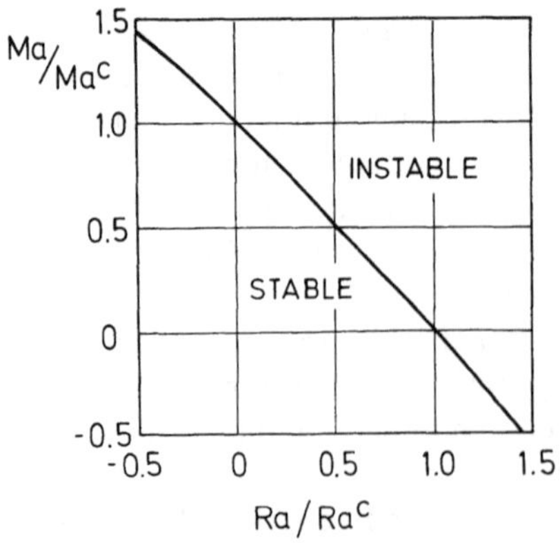

Fig. 18. Stability diagramm for the transition of "no flow" to convection cells of a fluid with free upper surface which is heated from below (after Nield[103])

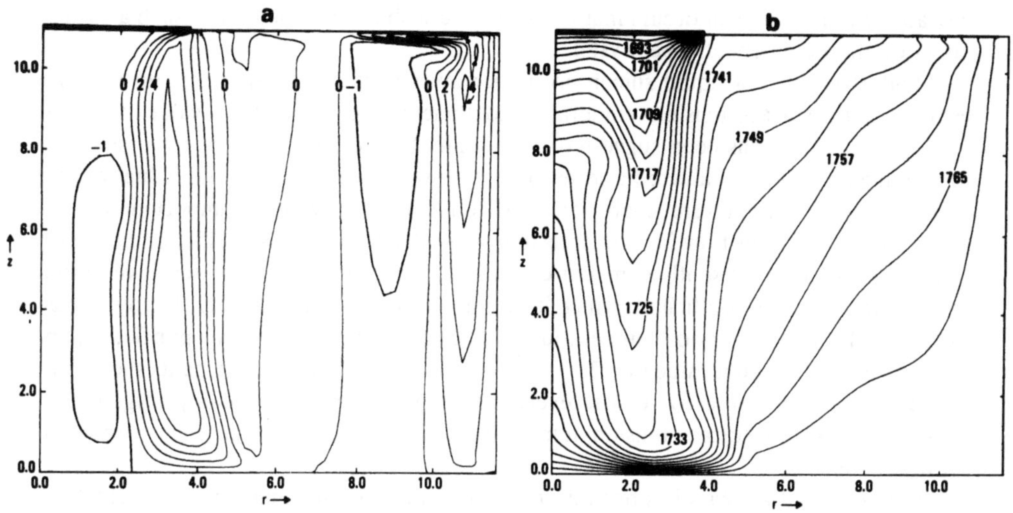

Fig. 19 a, b. The flow field of silicon in the right half of the crucible: **(a)** Streamlines, **(b)** Isotherms (after Langlois[110])

pattern[107]. It has been shown in this simulation that the spokes pattern is not a bulk-flow phenomenon and that it penetrates less than 1 cm beneath the surface[108]. In order to gain an understanding of the cause of the observed flow patterns one must consider both buoyancy driven and surface-tension-driven convection as has been done above.

Langlois[110] recently computed the flow in Czochralski growth of silicon crystals. The calculations were performed for a crucible radius of 11.55 cm, a crystal radius of 3.75 cm and a melt depth of 10.91 cm. A crucible rotation rate of 1.57 radians/sec and a crystal rotation rate (counter rotation) of −2.31 radians/sec were taken into account. The flow field and isotherms are shown in Fig. 19 which reveals the following features of the meridional circulation[110]:

1. An intense eddy, driven by thermocapillarity. The surface flow is rapid motion inward from the crucible wall. The return flow begins as a shallow layer but then forms a Taylor column along the crucible wall.
2. A weak back-eddy, driven by the return flow of the thermocapillarity circulation.
3. A broad region of near stagnation.
4. A columnar eddy driven mainly by crucible rotation, with some help from buoyant convection. The intensity of this eddy is reflected in the temperature plot, which shows isotherms displaced significantly downward by advective heat transfer in this region.
5. A weak but deep eddy generated by crystal rotation.

The numerical simulation was then continued with the surface-tension temperature coefficient set to zero, i.e. with thermocapillarity artificially turned off[110]. This is shown in Fig. 20. With no capillarity, the region of near stagnation is the entire annulus of melt lying under the free surface. Somewhat surprisingly, the centrifugal eddies are significantly modified. The flow generated by crystal rotation is compressed into a narrow boundary layer under the crystal face. The eddy generated by crucible rotation and buoyancy has expanded to fill the space, and has become weaker. The downward displacement of isotherms is correspondingly less pronounced. It is remarkable that thermocapillarity flow, which occurs mainly near the crucible wall, could have such a strong effect on the flow beneath the crystal.

Fluid properties of liquid silicon are under investigation. In particular, no reliable estimate of the surface-tension temperature coefficient is yet available. A value of $-4.3 \cdot 10^{-4}\ \mathrm{N \cdot m^{-1} \cdot K^{-1}}$ has been suggested[9], but might be too high, perhaps by a factor of four. Langlois[111] has conducted a series of digital simulations with various values of the surface-tension temperature coefficient and found that its value is important to the understanding of melt flow in Czochralski growth of silicon. If this coefficient lies near the upper end of its presently conjectured range, the flow would also be sensitive to the value of the viscosity coefficient, which itself remains uncertain. Recently, Hardy[112] has measured the surface-tension temperature coefficient of liquid silicon in a purified argon atmosphere and found a value of $-2.8 \cdot 10^{-4}\ \mathrm{N \cdot m^{-1} \cdot K^{-1}}$. His measurements show that the surface tension is sensitive to low levels of an impurity which is probably oxygen, and thus the measured surface-tension temperature coefficient might be higher than in growth systems.

A dimensional analysis has been performed by Balasubramaniam and Ostrach[113] to obtain the appropriate dimensionless parameters that describe the Czochralski bulk flow. Calculations for a representative case of silicon growth indicate that surface-tension-driven flow is the dominant flow mechanism. This is in accord with the more detailed results of numerical calculations by Langlois[110].

The oxygen content of Si crystals grown with the Czochralski technique from SiO_2 crucibles is an important specification for wafer processing. The SiO_2 crucible is slowly dissolved by the melt which becomes oxygen rich. A part of the oxygen is evaporated from the free melt surface as SiO vapor. The transport rate of oxygen from the melt to the atmosphere should depend on the fluid flow near the crucible wall and the flow near the free surface. In these regions surface-tension-driven flow plays a major rôle and should be considered in discussing the oxygen incorporation mechanism during Czochrolski growth.

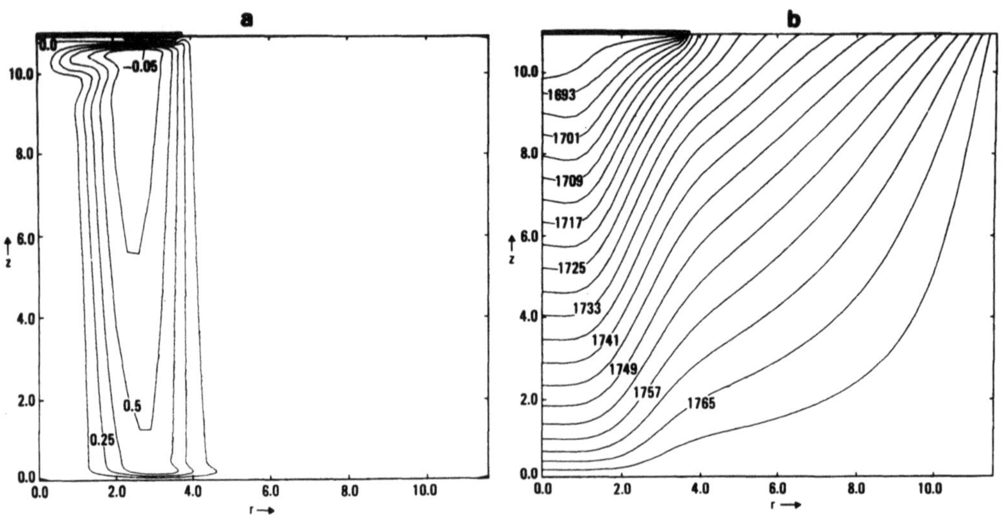

Fig. 20 a, b. The flow field of silicon in the right half of the crucible with thermocapillary turned off: **(a)** Streamlines, **(b)** Isotherms (after Langlois[110])

3.3 Surface-Tension-Driven Flow in Open-Boat Configurations

In open-boat zoning the liquid zone with a free upper surface is contained in a boat or a tube. The liquid zone is moved through the solid (e.g. by a travelling heater). The process is used for purification (especially metals and organic compounds) and thus fluid flow and mass transport in the liquid zone are important. Moreover, the open boat geometry is suited for two-dimensional calculations and model experiments; the configuration of an "open-boat" in crystal growth is related to what is called "open rectangular cavity" or "two-dimensional slot" in hydrodynamics.

In experiments with $NaNO_3$ zones in a rectangular open boat Schwabe and Scharmann[114, 115] could clearly demonstrate thermocapillary convection (Fig. 21), and they measured the velocity profile perpendicular to the free surface (Fig. 22), and the dependence of the surface velocity on the temperature difference. The surface velocity was found to increase linearily with the temperature difference ΔT. The boundary layer of thermocapillary convection shows itself in the steep velocity gradient near the free surface (Fig. 22). The authors could suppress thermocapillary forces with a film of viscous high-temperature oil. The flow velocity with the oil film was zero in the surface, and in the bulk liquid it was slow compared to the flow velocity with a free and clean surface. The authors[114)] noticed that the thermal conductivity of the side walls (e.g. of the crucible in the Czochralski technique) has to be high compared to the thermal conductivity of the

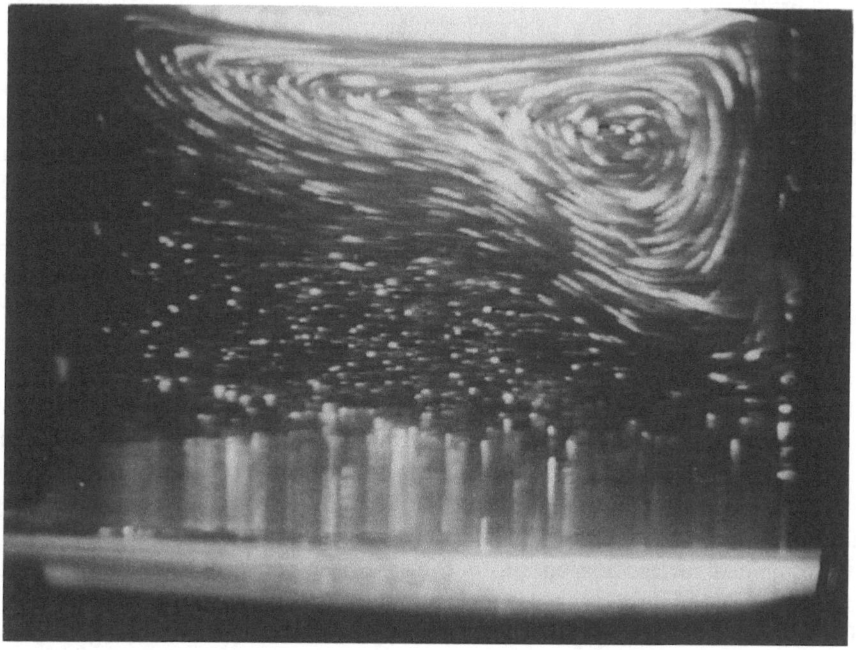

Fig. 21. Thermocapillary convection in a $NaNO_3$ molten zone (length L = 20 mm, width W = 20 mm, height H = 12 mm) between two graphite heating blocks (T_{left} = 380 °C, T_{right} = 330 °C) in a rectangular open boat. Photograph with ⅛ s (from a video monitor) observing a vertical light cut. The free surface is concave towards the air. Small bubbles are sticking to the vertical walls of the heating blocks (after Schwabe and Scharmann[114)])

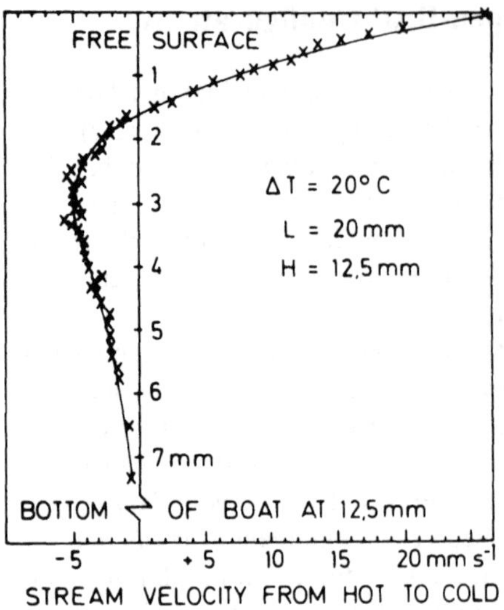

Fig. 22. Variation with depth of the horizontal stream velocity of thermocapillary convection in an open boat (after Schwabe and Scharmann[114])

melt in order to have strong thermocapillary flow. This is the case in the experiment with heated side walls made from graphite, and it also holds for real crystal growth of oxide or fluoride crystals from crucibles made of noble metals. Numerical calculations of the flow caused by thermocapillary forces at the free surface in $NaNO_3$ and Si melts in an open-boat geometry have been carried out by Wilke and Löser[116]. The calculated plane surface-tension-driven flow in a rectangular geometry agrees well even quantitatively with the measured flow velocities in the $NaNO_3$ melt. The relatively steep velocity gradient near the surface (Fig. 22), typical for surface-tension-driven flow, is reproduced. For the Si melt, which considerably differs in Prandtl number from the $NaNO_3$ melt, the same flow features were observed for small driving forces. For larger Marangoni numbers the flow has a more complicated multiple-eddy structure. Engelman and Sani[117] calculated velocity vectors, streamlines and temperature contours for the experiment with $NaNO_3$ melt by Schwabe and Scharmann[114]. The experimental and numerical results are in excellent agreement. To resolve the relative importance of surface-tension forces versus buoyancy forces in the experiment by Schwabe and Scharmann[114], two additional simulations were performed[117]. In the first, the buoyancy term in the momentum equation was neglected, while in the second the surface of the liquid was free and surface-tension effects were ignored. The results of the first case were nearly identical to those where both surface-tension and buoyancy forces are taken into account, whereas in the second case velocities were computed which are three orders of magnitude less than in the first case. This clearly indicates that the surface-tension effect is the driving force for this flow in the experiment.

Combined surface-tension and buoyancy-driven convection of silicone oils in two-dimensional configurations has been investigated experimentally and numerically[118, 119]. It was found[119] that the flow and temperature fields of the Marangoni convection strongly depend on the coupled buoyant convection and the aspect ratio of the liquid layer. Thin wires placed on the free surface (stretched in the direction perpendicular to the flow) were found to reduce the convective motion considerably. It was shown by calculations that the convection in electrically conducting melts is reduced greatly by applying a magnetic field perpendicular to the direction of the main stream.

Lamprecht, Schwabe and Scharmann[120)] studied thermocapillary and buoyant convection in an open cavity (boat) under normal and reduced gravity. The fluid was molten tetracosane ($C_{24}H_{50}$) in a 20 mm × 20 mm × 20 mm cubic cavity with free upper surface. The heating was from the side with the temperature gradient parallel to the free surface. Strong thermocapillary convection has been observed under reduced gravity when the surface was clean (Figs. 23–25). No flow was observed under milligravity when the free surface was covered with a dirt film. Such a dirt film can suppress thermocapillary forces totally as discussed previously for silicon oxide on silicon melt[83)]. Therefore, to have defined and reproducible flow conditions in the crystal-growth melt under reduced or normal gravity, one has to provide defined and reproducible chemical surface properties of the melt. This is a real challenge for crystal-growth systems and one which has not yet been met in most cases. The velocity profiles (Fig. 25) under normal and reduced gravity show that under microgravity we have one convection roll mixing the whole melt, whereas under normal gravity there is a convection roll with surface streaming and back streaming near the free surface, underneath wich there is almost no flow. This is also shown in Fig. 26 which should be compared with Fig. 24. One can see in Fig. 26 a convection roll driven by surface-tension forces near the free surface and a convection roll driven by buoyancy forces streaming up near the hot wall and down near the cold wall

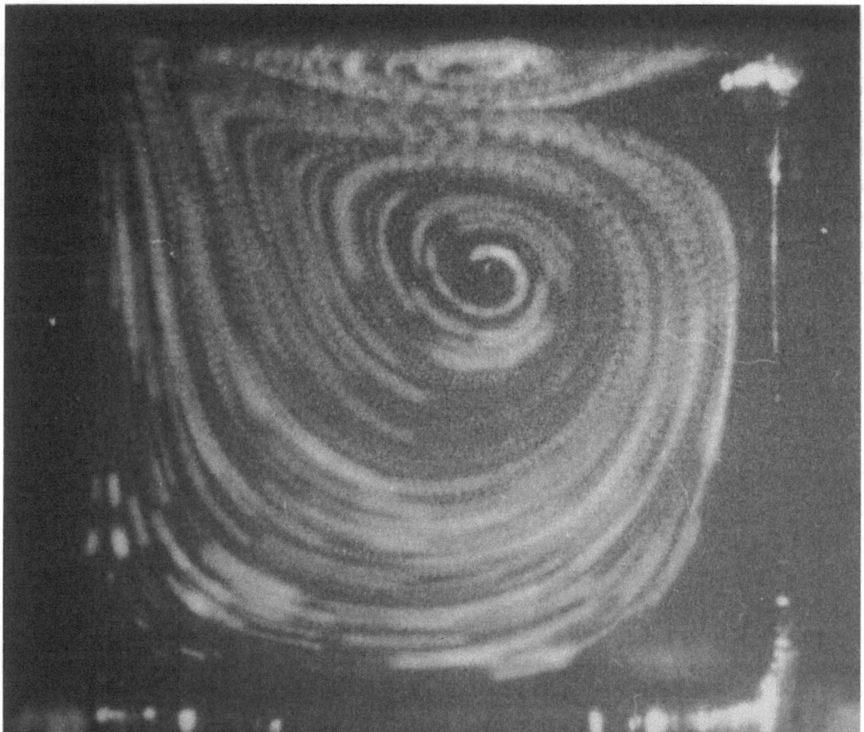

Fig. 23. Photograph of the streamlines of thermocapillary convection under reduced gravity (during the Spacelab D1-mission) with a temperature difference of 60 °C between the heaters. In front of the cold right side was dirt on the free surface (non-slip condition), suppressing thermocapillary flow in this region (after Lamprecht et al.[120)])

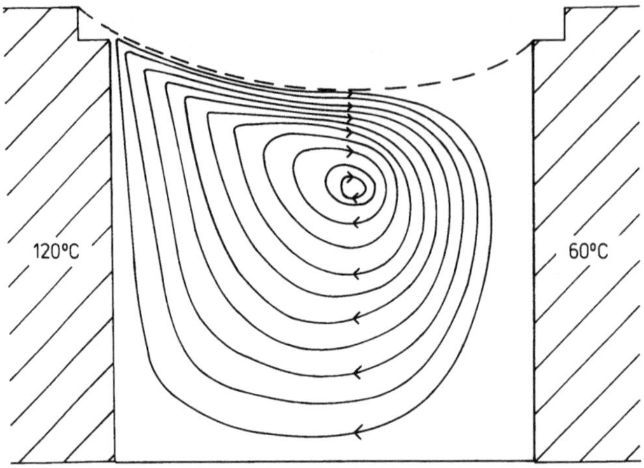

Fig. 24. Sketch of the streamlines of thermocapillary convection under reduced gravity drawn after the photograph in Fig. 23 (after Lamprecht et al.[120])

and along the bottom of the cavity. There is not much motion in the middle of the lower half of the cavity. This shows that the areas of predominantly surface-tension-driven flow and those of buoyancy-driven flow separate for large Marangoni numbers (Ma = 4.6×10^5, Gr = 3.3×10^5). Though the convection rolls of thermocapillary convection and of buoyant convection have the same sense of rotation, they do not combine into a large common convection roll, but, the areas separate and the rolls hinder each other. As seen in Fig. 25, the flow velocity in the free surface is larger under milligravity than under normal gravity. This shows that buoyant convection has a diminishing effect on the expansion and flow velocity of the thermocapillary convection roll. The heat and mass transport by pure thermocapillary convection under reduced gravity seem to be more effective than the coupled convection under normal gravity.

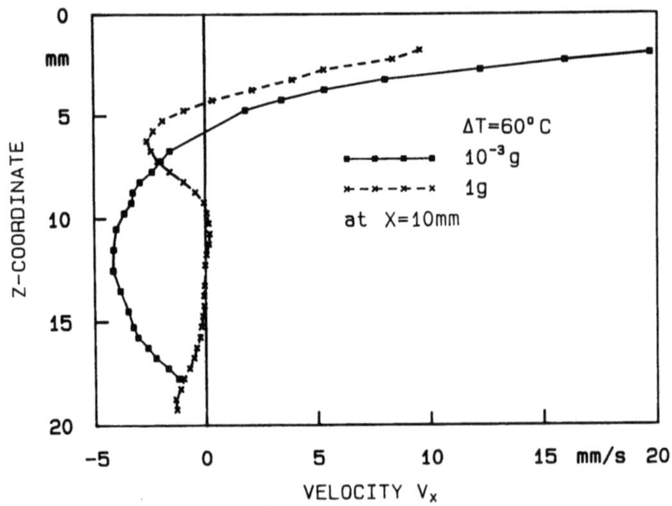

Fig. 25. Velocity profiles (variation with depth) of the flow parallel to the free surface from hot to cold (v_x) in the middle of the cuvette (at x = 10 mm) for a temperature difference of 60 °C between the two heaters for milligravity and normal gravity (after Lamprecht et al.[120])

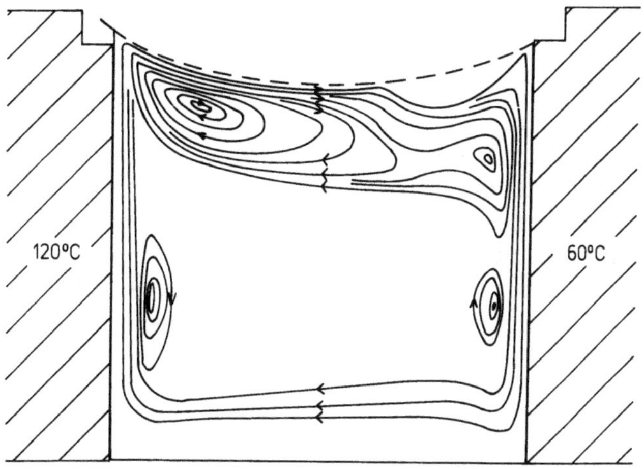

Fig. 26. Sketch of the streamlines of coupled thermocapillary and buoyant convection under normal gravity with a temperature difference of 60° C between the heaters (after Lamprecht et al.[120])

The surface tension σ is normally a decreasing function of temperature. However, some systems exhibit a special behavior where σ shows a minimum at a particular temperature T_{min}. This is the case for aqueous solutions of fatty alcohols[121] and some liquid alloys such as Ag-Pb and Al-Sn[122]. For temperatures T with $T < T_{min}$, we expect surface motion as usual from hot to cold but for $T_{min} < T$ the surface motion should be from cold to hot. Limbourg-Fontaine, Pétré and Legros[123] conducted such an experiment with n-heptanol aqueous solution which has a minimum in its surface tension versus temperatures curve. The experiment was performed during the free flight of a high-altitude research rocket (under reduced gravity). Thermocapillary convection occurred from the cold to the hot region in the liquid-gas interface of an n-heptanol solution submitted to a thermal gradient in the range where the surface tension is increasing with temperature. When the experiment was performed under normal gravity, surface movement from the cold to the hot wall was observed, with buoyancy-induced movements taking place in which the liquid rose up along the hot side and descended along the cold side. Such a fluid system with a minimum of surface tension was studied numerically by Villers and Platten[124]. When the surface tension increases linearly with temperature, the surface-tension forces pull a fluid surface element from the cold wall to the hot wall, and thus, this fluid element has to sink near the hot wall. However, due to gravity, a fluid element near the bottom must rise near the hot wall. Due to the combination of these two effects, two superposed counter-rotating cells are predicted by the calculations. Unfortunately, the computing scheme was stable only for small Marangoni and Grashof numbers so that the calculations cannot be directly compared with the experiment which was performed for much larger Ma and Gr.

Sen and Davis[125] and Strani, Piva and Graziani[126] consider thermocapillary convection in two-dimensional rectangular slots under zero gravity. Their results are valid only for thin liquid layers and small Marangoni numbers[125] or moderate Marangoni numbers[126]. Boundary layers of the temperature have been found to develop with increasing Marangoni number, and corresponding peaks of the surface velocity near the hot and cold wall are predicted[126]. For liquid whose depth is five times its length, the surface stream and its back stream are only significant in the upper part of the slot, showing the surface character of the phenomenon[126].

Steady thermocapillary flows in a square cavity with a free top surface in the absence of gravitational forces were computed by a finite-difference procedure[127]. The Prandtl number Pr was varied between 0.005 and 50, and the largest Marangoni number Ma attained with reasonable accuracy was 5×10^4. This are values which can be compared to those in real experiments[120]. For Pr = 50 and large Ma, boundary layers of the surface temperature develop near the hot and the cold wall. The surface velocity is larger near the hot wall than almost everywhere on the free surface except in the readily developed thermal layer at the cold wall .The opposite is true for Pr < 1 and low Ma, where convection is expected to dominate near the cold wall. These findings for Pr = 50 agree with observations made on fluids with Pr = 9[114] and Pr = 49[120] under normal earth gravity, where the suface velocity was always highest near the hot wall.

There is an experimental difficulty which hinders the observation of the theoretically predicted maximum surface veloctiy near the cold wall[127]; this is the presence of surface impurities. Such impurities will be swamped by the flow to the cold side and will accumulate in front of the cold wall, reducing the flow velocity there. Such effects of surface contamination on thermocapillary flow are considered by Homsy and Meiburg[128] and Carpenter and Homsy[129].

The free surface will be deflected by capillary and thermocapillary forces and pressure. This is studied in Refs. 125–127, 130. For Pr = 50 a depression near the hot wall and an elevation near the cold wall are predicted.

There have only been a few papers on thermocapillary convection that deal with real crystal growth situations or that describe thermocapillary effects in experimental work. The effect of the natural and thermocapillary convection on the vortex configuration in the solution during growth of PbTe crystals by the travelling-heater method is considered by Cherepanova and Apanovich[131]. In their calculations the authors vary the degree of the solution-surface contact with the ampoule wall, i.e. the degree of slippage of the melt at a solid surface. Polezhayev and coworkers[132] calculate streamlines and isotherms in a fluid with Prandtl number Pr = 1 under zero gravity.

The present author has collected scattered hints from the crystal-growth literature on open-boat zoning which, although not discussed in these terms, might appear to be connected with surface-tension-driven flow effects.

The interface during zoning of Pb indicated stronger convection in an open graphite boat (free melt surface) then in a closed glass tube (no free melt surface)[133]. Free slip at the free surface and additional thermocapillary forces might be the reason for this effect.

The main convection roll in a NaCl melt in a RF-heated open graphite boat (length 10.15 cm, height 1.27 cm and width 1.9 cm, temperature gradient along the free surface ≈ 30 K cm^{-1}, vertical temperature gradient ≈ 50 K cm^{-1}) was from hot to cold in the free surface with a velocity of ≈ 2.5 cm s^{-1}; the main flow carried some extra eddies, which gave rise to time-correlated temperature fluctuations at distant points along the main flow direction[134]. The boundary conditions give Ra $\approx 5 \cdot 10^4$ and Ma $\approx 2.5 \cdot 10^4$. Therefore, the limit for the onset of the cellular instability in a heated-from-below situation, $Ma/Ma^c + Ra/Ra^c \leq 1$ ($Ma^c \approx 70$ is the critical Marangoni number, $Ra^c = 1700$ is the critical Rayleigh number) is exceeded by buoyancy forces alone and by thermocapillary forces alone (see Fig. 18)[11]. the Bénard-Rayleigh instability, driven by the combined buoyancy and thermocapillary forces, must have existed in this experiment with the convection cells drifting in the main flow.

Temperature fluctuations in tin in an aluminum boat are pronounced near the free surface (in the upper most 1 cm of the 6.4 cm deep mould) and vanish when the surface is covered[135]. The non-slip condition, which applies when the surface is covered, suppresses thermocapillary forces, and at the same time reduces the buoyant convection; this makes differentiation between the two mechanisms difficult. The same increase in the amplitude of temperature oscillations towards the free surface has been found in another experiment with tin in an open boat[136]. The consideration of thermocapillary effects could shed new light on the experiments of Hurle and co-workers[137–139] who found regular temperature oscillations in a Ga melt with free surface subjected to a horizontal temperature gradient, and on the experiment of Carruthers and Pavilonis[140] who found temperature oscillations in a cylindrical $NaNO_3$ melt volume with a free surface, heated radially and cooled axially.

3.4 Further Methods and Problems in Crystal Growth Discussed in Connection with Thermocapillarity

3.4.1 Thermocapillarity in Ribbon Growth

Ribbon growth is a modern method for the production of shaped crystals (bands = ribbons, tubes etc.). Ribbon growth has been described by various authors[141]. The crystal is drawn from a shaper (die). The height of the melt meniscus between die and crystal interface is usually only about 1 mm or less. The free meniscus surface is subject to extreme temperature gradients of the order of 200 K cm^{-1} (heated from below, cooled from the side and from above). In Si-ribbon growth the Rayleigh number Ra is typically of the order of 1 and the Marangoni number M is of the order of 10^3. Ra is too small for the onset of convection in a heating-from-below situation and natural buoyant convection can occur because of horizontal gradients (cooling from the side). Thermocapillary convection sets in for $M > 0$, and the value $M \approx 10^3$ even exceeds the critical value for the onset of oscillatory thermocapillary convection as measured in $NaNO_3$ floating zones. Therefore strong thermocapillary convection is expected in the fluid meniscus and even oscillatory thermocapillary convection is possible.

Si ribbons grown under proper control of the growth atmosphere have the shiny metallic grey color of polished Si wafers[142]. Under certain lighting conditions, the ribbons appear to be covered by a multicolored film. This has been demonstrated to be a diffraction grating effect caused by transverse growth striations[142]. It is most likely that surface oscillations of the meniscus due to oscillatory thermocapillary convection are "frozen" into the ribbon. The surface-growth striations are very regular, do not change appreciably with distance along the ribbon, and increase in frequency roughly in proportion to the growth rate. This growth-rate dependence of the striation frequency is an indication that the surface modulation is not due to the pick up of external vibrations. Such surface striations have been observed to be "in phase" on either side of the thin ribbon, i.e., the ribbon thickness increases on both sides simultaneously[143].

The mechanism for lateral solute segregation in edge-defined film-fed crystal growth of Si ribbons was studied by Ettouney and Brown[144], taking thermocapillary convection into account. Thermocapillary forces cause intense laminar convection throughout the melt meniscus. The segregation patterns predicted with this flow always have maxima at the center of the ribbon and have concentration differences that are a factor of three or more larger than those predicted without this driving force. But this type of segregation inhomogeneity is not reported in the literature on real growth.

During growth of filamentary sapphire by the edge-defined film-fed growth technique, considerable turbulence was observed in the melt meniscus[145]. Periodic microvoid layers with a frequency of the order of 16 Hz are found in the filament. Examination of a 16 mm motion picture film, taken when the filament was growing, shows that the appearance of the void layer immediately above the meniscus, is preceded by a radial movement across the liquid-solid interface towards the center of the filament. This flow direction is in accord with that of thermocapillary flow. Obviously oscillatory thermocapillary flow in the melt meniscus is the reason for the oscillatory void inclusions. Thermocapillary flow in the Al_2O_3 melt meniscus should result in an interface which is convex towards the melt. This has also been observed[145, 146].

3.4.2 Thermocapillarity in Laser Melting

As proposed by Anthony and Cline[147], during laser surface melting and alloying, temperature gradients on the melt surface between the laser beam impact point and the intersection line of the solid-liquid interface with the surface, generate surface tension gradients that sweep liquid away from the beam impact point. The resulting flow of liquid creates a depression of the liquid surface beneath the beam and ridging of the liquid surface elsewhere. As the beam passes to other areas of the surface, this distortion of the liquid surface is frozen in, creating a roughened rippled surface. The authors could observe such rippling on an electron-beam-scanned surface of stainless steel. Cline[148] calculated and measured the surface rippling of a 2 μm thick film consisting of a 0.56 μm thick lead layer underlying a 1.44 μm thick tin layer on a Pyrex microscope slide. Alloying of lead and tin was observed by microprobe analysis after laser melting. The time necessary to alloy the 2 μm thick Pb-Sn film by liquid diffusion using a diffusion constant of 10^{-5} cm²/s is 4×10^{-3} s, which is much longer than the applied-heating time of the scanning laser. In contrast, fluid flow, which occurs first in the tin layer, reaches the Pb layer in 6×10^{-5} s. Tin flows away from the center of the laser beam and the surface becomes depressed. According to the calculations based on surface-tension gradients, the depression reaches the interface between the Sn and Pb layers during the 1.5×10^{-4} s of heating by the scanning laser. Once the surface contacts the lead liquid at the center of the depression, the surface tension changes from that of liquid tin to that of liquid lead. The surface tension of liquid lead is less than that of tin so the liquid lead would flow rapidly over the surface. The difference between the surface tension of tin and lead produces a larger force to drive the fluid flow than the thermal surface-tension gradient of pure tin alone. Thus, by considering surface-tension-driven flow, alloying of the lead and tin layers can be understood. Lasky[149] has investigated the effects of various encapsulating films on laser recrystallization of silicon on silicon oxide. Surface-tension gradients are eliminated as a driving force for the liquid silicon motion by such capping layers. A 4.0 nm thick film of SiO_2 is sufficient to accomplish this. Such an oxide capping had grown during laser recrystallization in air of nominally uncapped samples.

Laser-melted thin metal films on silicon show a cellular distribution of the metal. The cells, with the size of the order of the liquid-layer thickness, are proposed to be due to cellular convection forming Bénard cells[150, 151].

3.4.3 Thermocapillarity in Bubble Motion

Bubble inclusions are frequent and severe defects in crystals grown from the melt. Enormous efforts are undertaken to produce bubble-free crystals and glasses for optical applications. The solidification process must be managed such that bubbles, always present in the melt, are not captured by the solidification front. Thermocapillary convection can play a significant rôle in bubble motion and bubble capturing as will be discussed briefly in the following.

Bubbles in melts are small spheres with a free surface. A temperature (or concentration) gradient in the melt also exists near the bubble and induces thermocapillary convection (Fig. 27). This convection is part of the overall convection and can be of local importance for the wall or crystal interface to which the bubble sticks. On the other hand, thermocapillary convection around the bubble exerts a force on the buble propelling it from cold to hot regions. This bubble motion has been studied experimentally and theoretically[152–158]. When the viscosity of the gas in the bubble and the distortion of the temperature field due to bubble motion are neglected, a bubble moves under weightlessness from a colder to a hotter region with a velocity $v = -\frac{2}{3} \cdot \frac{R}{2\eta} \cdot \frac{\partial \sigma}{\partial T} \cdot \frac{\partial T}{\partial x}$, with bubble radius R and temperature gradient $\frac{\partial T}{\partial x}$. In case of concentration differences in the fluid, the bubble moves in the direction of decreasing surface tension.

Under normal gravity the forces of buoyancy on the bubble scale as R^3 whereas the surface-tension forces scale as R^2. Therefore buoyancy dominates for larger bubbles but for small bubbles the surface-tension forces can be dominant. Taking typical values from an oxide melt in Czochralski growth, the surface-tension forces are larger than the buoyancy forces for a bubble radius R of the order of 50 µm, which is a typical value for bubbles found in oxide crystals[159]. For the situation as encountered in Czochralski growth, buoyancy moves the bubbles upwards, which means towards the growing crystal interface, whereas thermocapillarity drives the bubbles away from the growing crystal. Thus thermocapillarity helps to avoid bubble inclusion. But the mechanism of bubble growth and bubble entrapment are very complicated and chemocapillary forces in particular must be taken into account. In front of the growing crystal a diffusion boundary layer of impurities exists with a steep concentration gradient. If the impurity lowers the

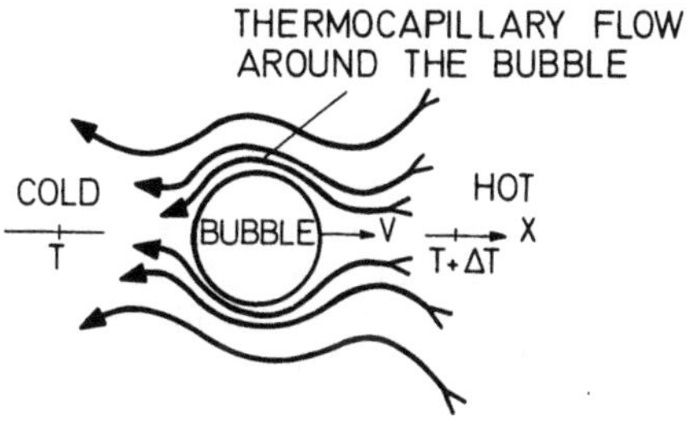

Fig. 27. A gas bubble in a fluid is propeled from cold to hot regions by thermocapillary flow around it

surface tension of the melt, chemocapillarity drives the bubbles towards the growing interface. Such chemocapillary bubble motion has not been observed in crystal growth till now, but impurities can also hinder thermocapillary bubble motion. Bubble motion in a temperature gradient in front of a growing crystal under microgravity has not been experimentally observed due to impurities[160, 161].

4 Conclusions and Future Outlook

The free melt surface and the temperature gradients inherent to many melt growth techniques generate thermocapillary forces which take part in driving fluid flow. This driving force becomes especially important in crystal growth under microgravity where buoyancy forces are absent.

Calculations and experiments show that surface-tension-driven flow has an influence on the crystal-growth conditions also in a laboratory under normal earth gravity. But it is difficult to ascertain the extent of this influence because thermocapillary convection is always coupled with buoyant convection under normal earth gravity. Only in small floating zones or in the melt meniscus of a ribbon can one have dominant thermocapillary convection with unequivocal effects on the grown crystal. Experiments under microgravity with the observation of pure thermocapillary convection and its impact on crystal growth can help to understand the rôle of thermocapillary convection in crystal growth on earth.

An estimate of the conditions in melt growth show that oscillatory thermocapillary convection can occur. Temperature oscillations due to oscillatory thermocapillary convection have been observed in model experiments and in real growth. This time dependence has not yet been studied in sufficient detail.

It is of prime importance in experiments on thermocapillary convection and in melt growth, to control the chemical conditions at the free surface since these determine the surface-tension properties. Thin layers of dirt can suppress thermocapillary forces totally and can strongly alter the fluid flow. As much attention should be paid to the state of the free melt surface as to the purity of the starting material.

Surface-tension-driven flow during the growth from a solution can be more complicated than the flow in nominally pure melts discussed thus far. Temperature gradients drive the flow from hot to cold in the free surface, but the solute, built up in front of the solidification interface, can change this flow direction if the solute lowers the surface tension. This has been observed during the solidification of tin-bismuth alloys[162].

Further interesting observations are expected to be reported from crystal growth systems if the crystal grower pays attention to surface-tension flow effects.

Acknowledgements. The work of the author on surface-tension-driven flow has been supported by the Federal Ministry for Research and Technology (BMFT) under contracts No. 01 QV 148-ZA-SN-SLN 7786, No. 01 QV-148, No. 01 QV 470 A1 and No. 01QV 064 A/8. This paper is dedicated to Prof. Dr. Drs. h.c. A. Scharmann, Justus-Liebig-Universität Gießen, on the occasion of his 60th birthday.

5 References

1. Hurle, D. T. J.: Hydrodynamics in Crystal Growth, in: Crystal Growth and Materials (eds. Kaldis, E., Scheel, H. J.) P. 549 North Holland 1977
2. Scriven, L. E., Sternling, C. V.: Nature *187,* 186 (1960)
3. Levich, V. G.: Physico-Chemical Hydrodynamics, New York, Englewood Cliffs 1962
4. Berg, J. C., Acrivos, A., Boudart, M.: Evaporative Convection, in: Adv. Chem. Eng. *6,* 61 (1966)
5. Kenning, D. B. R.: Appl. Mech. Rev. *21,* 1101 (1968)
6. Normand, Ch., Pomeau, Y., Velarde, M. G.: Rev. mod. Phys. *49,* 581 (1977)
7. Sørensen, T. S. (Ed.): Dynamics and Instability of Fluid Interfaces, Lecture Notes in Physics 105, Berlin, Heidelberg Springer 1979
8. Schwabe, D.: Physico Chem. Hydrodyn. *2,* 263 (1981)
9. Chang, Ch. E., Wilcox, W. R.: J. Crystal Growth *28,* 8 (1975)
10. Pearson, J. R. A.: J. Fluid Mech. *4,* 489 (1958)
11. Nield, D. A.: ibid. *19,* 341 (1964)
12. Ostrach, S.: Motion Induced by Capillarity, Proc. Int. Conf. Chem. Hydrodynamics, St. Peter Port, Guernsey Advance Publications 1978
13. Ostrach, S.: COSPAR Space Research *19,* 563 (1979)
14. Keller, W., Mühlbauer, A.: Preparation and Properties of Solid State Materials 5. Floating-Zone Silicon, New York and Basel, Marcel Dekker 1981
15. Dietze, W., Keller, W., Mühlbauer, A.: Float-zone Grown Silicon, in: Crystals, Growth, Properties and Applications *5,* (ed. Grabmaier J.) P.1, Berlin, Heidelberg Springer-Verlag 1981
16. Chang, Ch. E., Wilcox, W. R.: Int. J. Heat Mass Transfer *19,* 355 (1976)
17. Clark, P. A., Wilcox, W. R.: J. Crystal Growth *50,* 461 (1980)
18. Schwabe, D., Scharmann, A., Preisser, F., Oeder, R.: ibid. *43,* 305 (1978)
19. Chun, Ch.-H., Wuest, W.: ESA – SP – *142,* 283 (1979)
20. Chun, Ch.-H.: Acta Astronaut. *7,* 479 (1980)
21. Chun, Ch.-H.: J. Crystal Growth *48,* 600 (1980)
22. Chun, Ch.-H., Wuest, W.: Adv. Space Res. *1,* 17 (1981)
23. Schwabe, D., Scharmann, A.: J. Crystal Growth *46,* 125 (1979)
24. Schwabe, D., Scharmann, A., Preisser, F.: ESA – SP – *142,* 327 (1979)
25. Schwabe, D., Preisser, F., Scharmann, A.: DGLR – Bericht 80-02, 323 (1980)
26. Mc Neil, Th. J., Cole, R., Subramanian, R. S., Surface-Tension-Driven Flow in Glass Melts and Model Fluids, in: Materials processing in the reduced gravity environment of Space (Ed. Rindone G. E.) P. 189, Amsterdam Elsevier 1982
27. Kamotani, Y., Ostrach, S., Lowry, S.: An Experimental Study of Heat-Induced Surface-Tension Driven Flow, ibid, P. 161
28. Chun, Ch.-H., Wuest, W.: Acta Astronaut. *9,* 225 (1982)
29. Schwabe, D., Preisser, F., Scharmann, A.: Z. Flugwiss. Weltraumforsch. *6,* 309 (1982)
30. Schwabe, D., Preisser, F., Scharmann, A.: Acta Astronaut. *9,* 265 (1982)
31. Schwabe, D., Scharmann, A., Preisser, F.: ibid. *9,* 183 (1982)
32. Monti, R., Napolitano, L. G., Russo, G.: ESA - SP – 191, 219 (1983)
33. Chun, Ch.-H., Wuest, W.: ESA – SP – 191, 205 (1983)
34. Schwabe, D., Scharmann, A.: ESA – SP – 191, 213 (1983)
35. Schwabe, D., Scharmann, A.: Adv. Space Res. *3,* 89 (1983)
36. Preisser, F., Schwabe, D., Scharmann, A.: J. Fluid Mech. *126,* 545 (1983)
37. Kamotani, Y., Ostrach, S., Vargas, M.: J. Crystal Growth *66,* 83 (1984)
38. Monti, R., Napolitano, L. G., Mannara, G.: ESA – SP – 222, 229 (1984)
39. Monti, R., Napolitano, L. G., Russo, G.: Acta Astronaut. *11,* 369 (1984)
40. Mc Neil, Th. J., Cole, R., Subramanian, R. S.: J. Colloid Interface Sci. *98,* 210 (1984)
41. Chun, Ch.-H.: Acta Astronaut. *11,* 227 (1984)
42. Chun, Ch.-H.: ESA – SP – 222, 271 (1984)
43. Schwabe, D., Scharmann, A.: ESA – SP – 222, 281 (1984)
44. Schwabe, D., Scharmann, A.: Adv. Space Res. *4,* 43 (1984)
45. Napolitano, L. G.: Science *225,* 197 (1984)

46. Ostrach, S., Kamotani, Y., Lai, C. L.: Physico Chem. Hydrodyn. *6,* 585 (1985)
47. Napolitano, L. G., Monti, R., Russo, G.: Earth-Orient. Applic. Space Technol. *5,* 69 (1985)
48. Mc Neil, Th. J., Cole, R., Subramanian, R. S.: J. Amer. Ceramic Soc. *68,* 254 (1985)
49. Schwabe, D., Scharmann, A.: Z. Flugwiss. Weltraumforsch. *9,* 21 (1985)
50. Fu, B.-I., Ostrach, S.: Numerical Solution of Thermocapillary Flows in Floating Zones, in: Transport Phenomena in Materials Processing. ASME, PED – Vol. 10, HTD – Vol. 29, 1983
51. Fowlis, W. W., Roberts, G. O.: J. Crystal Growth *74,* 301 (1986)
52. Kozhoukharowa, Zk., Slavchev, S.: ibid. *74,* 236 (1986)
53. Bauer, H. F.: Ingenieur Archiv *55,* 242 (1985)
54. Napolitano, L. G., Golia, C.: Acta Astronaut. *11,* 213 (1984)
55. Bauer, H. F.: ibid. *11,* 301 (1984)
56. Bauer, H. F.: AIAA Journal *22,* 421 (1984)
57. Kobayashi, N.: J. Crystal Growth *66,* 63 (1984)
58. Xu, J.-J., Davis, S. H.: Phys. Fluids *26,* 2880 (1983)
59. Cowley, S. J., Davis, S. H.: J. Fluid Mech. *135,* 175 (1983)
60. Bauer, H. F.: Z. Flugwiss. Weltraumforsch. *7,* 120 (1983)
61. Bauer, H. F.: Acta Mechanica *46,* 167 (1983)
62. Bauer, H. F.: ZAMM *62,* 471 (1982)
63. Bauer, H. F.: Z. Flugwiss. Weltraumforsch. *6,* 252 (1982)
64. Graziani, G., Strani, M., Piva, P.: Acta Astronaut. *9,* 231 (1982)
65. Da-Riva, I., Pereira, E. A.: ibid. *9,* 217 (1982)
66. Chun, Ch.-H.: BMFT – FB W 77 – 42 (1977)
67. Chun, Ch.-H., Wuest, W.: Acta Astronaut. *6,* 1073 (1979)
68. Wuest, W., Chun, Ch.-H.: COSPAR Space Research *19,* 559 (1979)
69. Smith, M. K., Davis, S. H.: J. Fluid Mech. *132,* 119 (1983)
70. Smith, M. K., Davis, S. H.: ibid. *132,* 145 (1983)
71. Xu, J.-J., Davis, S. H.: The Physics of Fluids *27,* 1102 (1984)
72. Mühlbauer, A., Erdmann, W., Keller, W.: J. Crystal Growth *64,* 529 (1983)
73. Vaněk, P., Kadečková, S., Jurisch, M., Löser, W.: ibid. *63,* 191 (1983)
74. Barthel, J., Eichler, K., Jurisch, M., Löser, W.: Kristall und Technik *14,* 637 (1979)
75. Jurisch, M., Löser, W., Lyumkis, E., Martuzane, E., Martuzans, B.: Crystal Res. & Technol. *17,* 963 (1982)
76. Kitamura, K., Ii, N., Shindo, I., Kimura, S.: J. Crystal Growth *46,* 277 (1979)
77. Kölker, H.: ibid. *50,* 852 (1980)
78. Kölker, H.: ESA – SP – 222, 169 (1984)
79. Kölker, H.: Private Communication (1986)
80. Eyer, A., Leiste, H., Nitsche, R.: ESA – SP – 222, 173 (1984)
81. Eyer, A., Leiste, H., Nitsche, R.: J. Crystal Growth *71,* 173 (1985)
82. Eyer, A., Leiste, H.: ibid. *71,* 249 (1985)
83. Croell, A., Mueller, W., Nitsche, R.: ibid. *79,* 65 (1986)
84. Witt, A. F., Gatos, H. C., Lichtensteiger, M., Lavine, M. C., Hermann, J. C.: J. Electrochem. Soc. *122,* 276 (1975)
85. Barmin, I. V., Zemskov, V. S., Raukhman, M. R., Sechenkov, A. S., Antipov, A. I., Egorov, A. V., Agapova, E. A.: IAF Congress Paper IAF – 80 – C – 103, Tokyo (1980)
86. Langlois, W. E.: Physico Chem. Hydrodyn. *2,* 245 (1981)
87. Pimputcar, S. M., Ostrach, S.: J. Crystal Growth *55,* 614 (1981)
88. Kobayashi, N.: ibid. *52,* 425 (1981)
89. Hurle, D. T. J.: Ibid. *65,* 124 (1983)
90. Langlois, W. E.: Ann. Rev. Fluid Mech. *17,* 191 (1985)
91. Langlois, W. E.: J. Crystal Growth *48,* 25 (1980)
92. Lamprecht, R., Schwabe, D., Scharmann, A., Schultheiss, E.: ibid. *65,* 143 (1983)
93. Carruthers, J. R.: ibid. *36,* 212 (1976)
94. Miyazawa, S.: ibid. *53,* 636 (1981)
95. Kobayashi, N.: ibid. *55,* 339 (1981)
96. Brandle, C. D.: ibid. *57,* 65 (1982)
97. Brice, J. C., Bruton, T. M., Hill, O. F., Whiffin, P. A. C.: ibid. *24/25*, 429 (1974)
98. Brandle, C. D., Miller, D. C.: ibid. *24/25*, 432 (1974)
99. Brandle, C. D.: ibid. *42,* 400 (1977)

100. Wilcox, W. R., Fullmer, L. D.: J. Appl. Phys. *36,* 2201 (1965)
101. Whiffin, P. A. C., Brice, J. C.: J. Crystal Growth *10,* 91 (1971)
102. Cockayne, B., Gates, M. P.: J. Mater. Sci. *2,* 118 (1967)
103. Nield, D. A.: J. Fluid Mech. *19,* 341 (1964)
104. Grodzka, P. G., Bannister, T. C.: Science *176,* 506 (1972)
105. Grodzka, P. G., Bannister, T. C.: ibid. *187,* 165 (1975)
106. Jones, A. D. W.: J. Crystal Growth *63,* 70 (1983)
107. Miller, D. C., Pernell, T. L.: ibid. *53,* 523 (1981)
108. Miller, D. C., Pernell, T. L.: ibid. *57,* 253 (1982)
109. Miller, D. C., The Role of Fluid Flow Phenomena in the Czochralski Growth of Oxides, in: Materials Processing in the Reduced Gravity Environment of Space (Ed. Rindone, G. E.) P. 373, Amsterdam, Elsevier 1982
110. Langlois, W. E.: Computer Meth. Appl. Mech. Engng. *25,* 315 (1981)
111. Langlois, W. E.: J. Crystal Growth *56,* 15 (1982)
112. Hardy, S. C.: ibid. *69,* 456 (1984)
113. Balasubramaniam, R., Ostrach, S.: Physico Chem. Hydrodyn. *5,* 3 (1984)
114. Schwabe, D., Scharmann, A.: J. Crystal Growth *52,* 435 (1981)
115. Schwabe, D., Scharmann, A.: Lett. Heat Mass Transfer *7,* 283 (1980)
116. Wilke, H., Löser W.: Crystal Res. and Technol. *18,* 825 (1983)
117. Engelman, M. S., Sani, R. L., Finite Element Simulation of Temperature-Dependent Free Surface Flows, in: Numerical Methods in Laminar and Turbulent Flow (Eds. Taylor, C., Olson, M. D., Gresho, P. M., Habashi, W. G.) P. 1325, Pineridge Press, Swansea, U. K. 1985
118. Sezaki, K., Enya, S., Morioka, M., Ochiai, J., Tanasawa, I., Maekawa, T.: Adv. Space Res. 3, 85 (1983)
119. Ochiai, J., Kuwahara, K., Morioka, M., Enya, S., Sezaki, K., Maekawa, T., Tanasawa, I.: ESA - SP – 222, 291 (1984)
120. Lamprecht, R., Schwabe, D., Scharmann, A.: J. Fluid Mech. submitted (1987)
121. Legros, J. C., Pétré, G., Limbourg-Fontaine, M. Cl.: ESA - SP – 191 (1983)
122. Desre, P. J., Joud, J. C.: Acta Astronautica *8,* 407 (1981)
123. Limbourg-Fontaine, M. C., Pétré, G., Legros, J. C.: Physico Chem. Hydrodynam. *6,* 301 (1985)
124. Villers, D., Platten, J. K.: ibid. *6,* 435 (1985)
125. Sen, A. K., Davis, S. H.: J. Fluid Mech. *121,* 163 (1982)
126. Strani, M., Piva, R., Graziani, G.: ibid. *130,* 347 (1983)
127. Zebib, A., Homsy, G., Meiburg, E.: Phys. Fluids *28,* 3467 (1985)
128. Homsy, G. M., Meiburg, E.: J. Fluid Mech. *139,* 443 (1984)
129. Carpenter, B., Homsy, G. M.: ibid. *155,* 429 (1985)
130. Cuvelier, C., Driessen, J. M.: ibid. *169,* 1 (1986)
131. Cherepanova, T. A., Apanovich, Y. V.: Crystal Res. & Technol. *18,* 287 (1983)
132. Polezhayev, V. I., Dubovik, K. G., Nikitin, S. A., Prostomolotov, A. I., Fedyushkin, A. I.: J. Crystal Growth *52,* 465 (1981)
133. Carruthers, J. R. Winegard, W. C.: Physics Chem. Solids, Suppl. No. 1, 645 (1967)
134. Utech, H. P., Brower, W. S., Early, J. G.: ibid. Suppl. No. 1, 201 (1967)
135. Stewart, J. M., Weinberg, F.: J. Crystal Growth *12,* 228 (1971)
136. Utech, H. P., Flemings, M. C.: Physics Chem. Solids, Suppl. No. 1, 651 (1967)
137. Hurle, D. T. J.: Phil. Mag. *13,* 305 (1966)
138. Hurle, D. T. J., Gillman, J., Harp, E. J.: ibid. *14,* 205 (1966)
139. Hurle, D. T. J., Jakeman, E., Johnson, C. P.: J. Fluid Mech. *64,* 565 (1974)
140. Carruthers, J. R., Pavilonis, J.: J. Appl. Phys. *39,* 5814 (1968)
141. Various authors: J. Crystal Growth *50,* 1 (1980)
142. Czisek, T. F., Schwuttke, G. H., Yang, K. H.: ibid. *50,* 160 (1980)
143. Surek, T., Hari Rao, C. B., Swartz, J. C., Garone, L. C.: J. Electrochem. Soc. *124,* 112 (1977)
144. Ettouney, H. M., Brown, R. A.: J. Appl. Phys. *55,* 4384 (1984)
145. Pollock, J. T. A.: J. Mater. Sci. *7,* 631 (1972)
146. Wada, K., Hoshikawa, K.: J. Crystal Growth *44,* 502 (1978)
147. Anthony, T. R., Cline, H. E.: J. Appl. Phys. *48,* 3888 (1977)
148. Cline, H. E.: ibid. *52,* 443 (1981)
149. Lasky, J. B.: ibid. *53,* 9038 (1982)

150. Possin, G. E., Parks, H. G., Chiang, S. W., Convection in Pulsed Laser Formed Melts, in: Laser and Electron – Beam Solid Interactions and Materials Processing (Eds. Gibbons, J. F., Hess, L. D., Sigmon, T. W.) P. 73, Amsterdam Elsevier, North Holland 1981
151. Van Gurp, G. J., Eggermont, G. E. J., Tamminga, Y., Stacy, W. T., Gijsbers, J. R. M.: Appl. Phys. Lett. *35,* 273 (1979)
152. Young, N. O., Goldstein, J. S., Block, M. J.: J. Fluid Mech. *6,* 350 (1959)
153. Harper, J. E., Moore, D. W., Pearson, J. R. A.: ibid. *27,* 361 (1967)
154. Hardy, S. C.: J. Colloid Interface Sci. *69,* 157 (1979)
155. Subramanian, R. S.: AIChE Journal *27,* 646 (1981)
156. Subramanian, R. S.: Adv. Space Res. *1,* 145 (1983)
157. Klein, H., Feuerbacher, B.: Wärme- Stoffübertragung *17,* 115 (1983)
158. Subramanian, R. S.: J. Fluid Mech. *153,* 389 (1985)
159. Cockayne, B.: J. Crystal Growth *42,* 413 (1977)
160. Papazian, J. M., Wilcox, W. R.: AIAA Journal *16,* 447 (1978)
161. Klein, H., Nähle, R., Wanders, K.: Z. Flugwiss. Weltraumforsch. *9,* 14 (1985)
162. Camel, D., Tison, P., Favier, J. J.: Meeting F.N.R.S. "Convection, Surface Tension, Microgravity", May 1986, Mons (Belgium)

Electrooptical Effects, Crystals and Devices

Hans-J. Weber

Universität Dortmund, Institut für Physik, 4600 Dortmund 50, F.R.G.

The propagation of light through an electrooptical crystal is described on basis of a phenomenological theory. Some physical aspects determining the magnitude of electrooptic effects are given and different experimental methods for measuring these effects are discussed. As representative examples of important electrooptical materials $LiNbO_3$, tetragonal bronzes, perovskites, KDP, and semiconductors with sphalerite structure are characterized with respect to crystal growth, preparation techniques and dielectric properties. The magnitude of their electrooptic coefficients is critically reviewed. Underlying ideas of different types of modulators and recent progress in fabrication of devices are discussed. These include bulk modulators, deflectors, and lumped and traveling wave modulators made from waveguides. Finally electrooptic bistability and the photorefractive effect which is the basis of holographic storage are explained. The present review may serve as an introduction to electrooptic effects, as a documentation of special applications of crystals and as a guide to the different scientific fields involved in this important branch of "high technology".

Crystals 11

1 Introduction

The first milestone in scientific research into electrooptical effects is represented by the work of Pockels. In his famous book "Lehrbuch der Kristalloptik", which dates from the year 1906, he established the phenomenological theory of the "Pockels-effect" in anisotropic materials. About 20 years ago a new interest in electrooptical effects was excited by the need for light modulators in laser technology. At first various concepts were developed and realized by using bulk crystals. Simultaneously the techniques in crystal growth were improved and new materials were found. An important step towards technical applications was the preparation of waveguides in the crystals. Due to their small dimensions the voltages could be decreased and the modulation frequencies could be increased drastically. Nowadays commercial devices with a bandwidth of several gigahertz are available.

In the last decade new effects have been investigated which extend potential applications of the electrooptical effect. Bistable devices can be used as elements with different functions and optical storage of information is made possible by the photorefractive effect. A new technique of crystal growth which will influence the concepts of device fabrication in future is the molecular beam epitaxy. With this tool in material science the optical properties of a sample can be taylored on a very small scale, which facilitates the fabrication of integrated optical devices.

Several decades elapsed from Pockels' electrooptical experiments until a technical device was developed. Nowadays Pockels-cells are widely used and the importance of the effect is still increasing. Its use in optical communication systems will be a very attractive and fascinating application of classical crystal optics in modern technology.

2 Crystal Optical Phenomena

2.1 Optical Properties of Crystals

Disregarding nonlinear effects, the relation between the dielectric displacement $\vec{D}$ and the electric field $\vec{E}$ of an optical wave is given by

$$D_i = \varepsilon_{ij}^* E_j + \gamma_{ijk}^* \nabla_k E_j \ . \tag{2.1}$$

Here and in subsequent equations we leave out the summation sign and follow the convention about repeated indices. In (2.1) ε_{ij}^* represents the complex dielectric and γ_{ijk}^* the complex gyration tensor. Their real parts, ε_{ij} and γ_{ijk}, describe the retardation and their imaginary parts, ε'_{ij} and γ'_{ijk}, the absorption of linearly and circularly polarized waves, respectively. The effects are called refraction (ε_{ij}), absorption (ε'_{ij}), optical activity or gyration (γ_{ijk}), and circular dichroism (γ'_{ijk}). γ_{ijk}^* is antisymmetric in its first two indices. Therefore it can be reduced to an axial tensor of rank two:

$$\gamma_{ijl}^* = e_{ijk} g_{kl}^* \ ,$$

e_{ijk} being the Levi-Civita tensor.

Now the constitutive Eqs. (2.1) have to be combined with Maxwell's equations. This procedure is described in several standard text books about crystal optics[1)]. The results are expressed in terms of the refractive index n, occuring in the expression for the wave number $|\vec{k}| = \frac{2\pi}{\lambda_0} n$ of the plane wave

$$\vec{D} = \vec{D}_0 \cdot e^{i(\omega t - \vec{k} \cdot \vec{z})} .$$

Here λ_0 denotes the wavelength in vacuum, ω the circular frequency and $\vec{z}$ the propagation direction. In materials with the magnetic permeability $\mu = 1$ and with $g^* = 0$ (nonmagnetic and nongyrotropic crystals) the index of refraction is given by

$$n^* = \sqrt{\varepsilon^*} = n - i\varkappa .$$

Traditionally, electric field induced refraction and absorption effects are described on the basis of a relation which is inverse to Eq. (2.1):

$$E_i = a_{ij}^* D_j .$$

The complex quantities

$$a_{ij}^* = a_{ij} + i a'_{ij}$$

are called polarization constants. The relation between a^* and n^* is expressed by

$$a = \frac{n^2 - \varkappa^2}{(n^2 - \varkappa^2)^2 + (2\,n\varkappa)^2} \tag{2.2a}$$

and

$$a' = \frac{2\,n\varkappa}{(n^2 - \varkappa^2)^2 + (2\,n\varkappa)^2} . \tag{2.2b}$$

Equation (2.2) is exactly valid only when the principal axes of the refraction and absorption tensor coincide with the coordination axes of the reference system. In this case two linearly polarized orthogonal waves transverse the crystal. Their indices of refraction are

$$n_{x,y} = \sqrt{\frac{1}{a_{x,y}}} \tag{2.3a}$$

and their absorption coefficients $\alpha = \frac{1}{L} \ln \frac{I_0}{I}$

(L: thickness of sample; I: intensity of light) are given by

$$\alpha_{x,y} = \frac{4\pi}{\lambda_0} \varkappa_{x,y} = \frac{2\pi}{\lambda_0} n^3 a'_{x,y} . \tag{2.3b}$$

In (2.3b) we have introduced the experimentally relevant approximation $\varkappa \ll 1$. The existence of two different refractive indices for one propagation direction of the light beam is called birefringence. It is one of the most important quantities in electrooptics and is expressed in terms of the phase difference of both waves as

$$\delta = \frac{2\pi}{\lambda_0}(n_y - n_x)L \; .$$

Optical activity manifests itselfs as a rotation ϱ of a linearly polarized optical wave:

$$\varrho = \frac{\pi L}{\lambda_0 n} g_{ij} l_i l_j \; ,$$

where l_i are components of the unit wave vector. The circular dichroism changes the linear polarization to an elliptical polarization by the amount

$$\vartheta = \frac{1}{2}\arctan(e^{2\sigma})$$

with

$$\sigma = \frac{\pi L}{\lambda_0 n} g'_{ij} l_i l_j \; .$$

Here g'_{ij} is the imaginary part of the gyration tensor. Using the concept of tensor surfaces[2)] the laws of crystal optics can be illustrated by the indicatrix

$$a_{ij} x_i x_j = 1 \tag{2.4}$$

which represents an ellipsoid. The lengths of its principal axes are proportional to the three principal refractive indices of a crystal.

When an optical wave propagates in an oblique direction through a crystal the two vibration directions $D_{x,y}$ and both indices of refraction $n_{x,y}$ are found by drawing the central section of the indicatrix perpendicular to the $\vec{k}$-vector. This will be an ellipse. The length and the orientation of its semi-axes determine $n_{x,y}$ and $D_{x,y}$, respectively. With a similar geometrical construction the orientation of Poynting's vector $\vec{s}$, which is the direction of energy flow, can be found. Generally $\vec{s}$ and $\vec{k}$ do not coincide, which is important in some devices as shown in Chap. 5. The geometrical illustrations of light propagation in birefringent crystals are shown in Fig. 2.1.

An ellipsoid, similar to (2.4) can be constructed with the absorption tensor:

$$a'_{ij} x_i x_j = 1 \; .$$

In crystals of low symmetry the orientation of both ellipsoids does not necessarily coincide. In this case the two principle waves traversing a crystal plate are elliptically polarized. Their polarization states are even no longer orthogonal, which results in additional interference phenomena of the emergent light[3)].

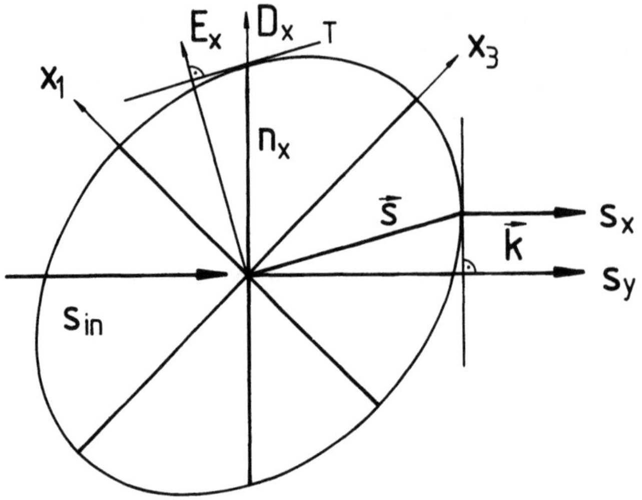

Fig. 2.1. Poinsot-construction for light propagation in a birefringent crystal[2]. The figure shows the principal section of the indicatrix of an uniaxial crystal with the optical axis parallel to x_3. D_x is the dielectric displacement vector of the extraordinary wave. The dielectric displacement of the ordinary wave, D_y, is standing perpendicular on the plane of drawing. s_x and s_y denote the directions of $\vec{k}$ and $\vec{s}$ outside the crystal for the extraordinary and the ordinary wave, respectively. T is a tangential plane for constructing E_x and the ray direction of D_x within the crystal. For s_{in} perpendicular to a planparallel plate of thickness L the spatial separation of both rays amounts $L \cdot \cos\zeta$, ζ being the angle between D_x and E_x

Severe difficulties arise with a birefringent, absorbing and gyrotropic crystal. Approximative methods describing light propagation have been developed by Pancharatnam[4] and Jones[5]. Recently Yariv and Lotspeich[6] have proposed a coupled-mode theory for birefringent and optically active materials. A proper test of its usefulness by extending the theory to absorbing crystals has not been realized until now.

2.2 Electrooptic Tensors

As the impact of an electric field on the optical properties is usually small, electrooptical effects are described as power series of a_{ij} and g_{ij}:

$$a_{ij} = a_{ij}^0 + r_{ijk}E_k + R_{ijkl}E_kE_l + \dots \quad \text{(a)}$$

$$a'_{ij} = a'^0_{ijk} + r'_{ijk}E_k + R'_{ijkl}E_kE_l + \dots \quad \text{(b)}$$

$$\text{(2.5)}$$

$$g_{ij} = g_{ij}^0 + s_{ijk}E_k + \dots \quad \text{(c)}$$

$$g'_{ij} = g'_{ij} + s'_{ijk}E_k + \dots \quad \text{(d)}$$

In Table 2.1 we present the nomenclature and some general features of the electrooptic tensors. The suffixes in (2.5) run from 1 to 3. In a contracted notation indices from 1 to 6 are used and the tensor components in (2.5) are changed accordingly to

$$a_{ij} \rightarrow a_m \qquad \gamma_{ij} \rightarrow \gamma_m \qquad (k = 1, 2, 3$$
$$r_{ijk} \rightarrow r_{mk} \qquad s_{ijk} \rightarrow s_{mk} \qquad m, n = 1, 2 \dots 6)$$
$$R_{ijkl} \rightarrow R_{mn}$$

In abbreviating the suffixes in R_{ijkl}, a factor of 2 appears:

$$R_{mn} = R_{ijkl}\,, \text{ when } n = 1, 2 \text{ or } 3$$
$$R_{mn} = 2\,R_{ijkl}\,, \text{ when } n = 4, 5 \text{ or } 6\,.$$

Textbooks dealing with tensor calculations in crystal physics are given in Ref. 2. Most commonly used is the book of Nye. Narasimhamurty considers extensively the photoelastic and the electrooptic effect and the most comprehensive description of physical properties of crystals is given by Haussühl.

2.3 Constraints by Symmetry

Generally a tensor of rank r possesses r^3 components in 3-dimensional space. Therefore the number of components for tensors of rank 2, 3, and 4 is 9, 27, and 81, respectively. As $a_{ij} = a_{ji}$, the first two indices of all electrooptic coefficients can be interchanged. The same is true for the last two indices of the quadratic effect: $R_{ijkl} = R_{ijlk}$. Thus we are left

Table 2.1. Optical and electrooptical tensors in crystals

Abbreviation	General relations	Nomenclature	Kind and rank	Number of independent components
a_{ij}	$a_{ij} = a_{ji} = a_m$	polarization constants or dielectric impermeability at optical frequencies	polar 2	6
a'_{ij}	$a'_{ij} = a'_{ji} = a'_m$	absorption tensor	polar 2	6
γ_{ij}	$\gamma_{ij} = \gamma_{ji} = \gamma_m$	optical activity or gyration	axial 2	6
γ'_{ij}	$\gamma'_{ij} = \gamma'_{ji} = \gamma'_m$	circular dichroism or Cotton-effect	axial 2	6
r_{ijk}	$r_{ijk} = r_{jik} = r_{mk}$	linear electrooptic effect or Pockels-effect	polar 3	18
r'_{ijk}	$r'_{ijk} = r'_{jik} = r'_{mk}$	linear electroabsorption	polar 3	18
s_{ijk}	$s_{ijk} = s_{jik} = s_{mk}$	electrogyration	axial 3	18
s'_{ijk}	$s'_{ijk} = s'_{jik} = s'_{mk}$	circular electrodichroism	axial 3	18
R_{ijkl}	$R_{ijkl} = R_{jikl} = R_{ijlk} = R_{mn}$	quadratic electrooptic or Kerr-effect	polar 4	36
R'_{ijkl}	$R'_{ijkl} = R'_{jikl} = R'_{ijlk} = R'_{mn}$	quadratic electroabsorption	polar 4	36

with 18 r_{ijk} and 36 R_{ijkl}, which are further reduced by symmetry elements of the crystals. Let α_{ij} be the matrix of a symmetry operation. Then the identity

$$r_{ijk} = \alpha_{il}\alpha_{jm}\alpha_{kn}r_{lmn}$$

holds for a polar tensor of rank 3. Equivalent relations are valid for tensors of higher order. For axial tensors the right side of the above equation has to be multiplied with D^r, D being the determinant of α_{ij}. $D = 1$, if the handness of the reference system is maintained and $D = -1$, if it is reversed by the symmetry operation. The most important operation affecting the existence of physical properties is a centre of symmetry. In centrosymmetric crystals polar tensors of odd rank (r_{ijk}) and axial tensors of even rank (g_{ij}) vanish. The matrices of electrooptic tensors in different point groups are considered in Chap. 4 as far as they are needed.

2.4 Secondary Effects

The electrooptic response coefficients contain different macroscopic contributions. In detail they depend on the conditions under which an experiment is carried out. Due to electromechanical and electrocaloric couplings secondary effects occur. They are taken into account by suitable extensions of Eq. (2.5). As an instructive example we consider the polarization constants (2.5a) as a function of electric fields E_i, mechanical stresses σ_{kl} and temperature changes ΔT:

$$\Delta a_{ij} = r_{ijk}^{\sigma,T}E_k + R_{ijkl}^{\sigma,T}E_kE_l + q_{ijkl}^{E,T}\sigma_{kl} + t_{ij}^{\sigma,E}\Delta T \qquad (2.6)$$

Here q_{ijkl} and t_{ij} represent the piezooptical tensor and the temperature dependence of the a_{ij}. The superscripts denote the thermodynamic conditions under which the coefficients have to be measured.

The stress in (2.6) appears as a consequence of the attraction which a pair of oppositely charged electrodes exerts. Assuming that the crystalline sample is formed as a plate condenser,

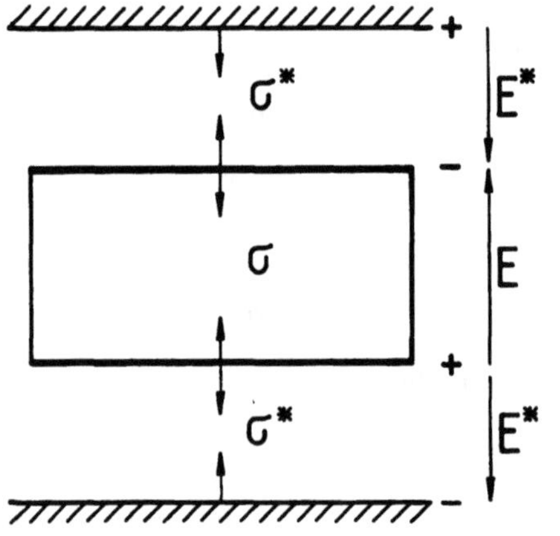

Fig. 2.2. Compensation of stresses by additional electrodes. σ* is the stress which is produced by the additional field E* and which compensates σ, the stress within the sample

$$\sigma = -(\varepsilon_0/2)\varepsilon E^2 , \tag{2.7}$$

where ε_0 and ε are the dielectric constants of the vacuum and the condenser material, respectively. Equation (2.7) shows that σ depends quadratically on $\vec{E}$. Therefore it occurs only in Kerr-effect measurements. Weber and Haussühl[7] have demonstrated that the stress can be controlled with electrodes charged and positioned as shown in Fig. 2.2.

The electric field induces temperature changes ΔT via the linear and quadratic electrocaloric effects. If ac-fields are applied usually no thermal equilibrium between the sample and its surrounding medium is attained within one oscillation and the measured coefficients are adiabatic ones. Employing standard thermodynamic methods[2] the relation between adiabatic and isothermal constants reads

$$r^T_{ijk} - r^S_{ijk} = t_{ij} p_k T/C$$

for the linear effect and

$$R^T_{ijkl} - R^S_{ijkl} = 0.5 \; t_{ij} \varepsilon_0 \frac{\partial \tilde{\varepsilon}_{kl}}{\partial T} \; T/C$$

for the quadratic effect. $\partial \tilde{\varepsilon}_{kl}/\partial T$ describes the temperature dependence of the relative dielectric constant and $t_{ij} = \partial a_{ij}/\partial T$ the changes of the optical polarization constant, C is the specific heat of the crystal and p_k its pyroelectric coefficient.

Even if the electrodes are free from stress, the electric field deforms the sample due to electrostrictive effects:

$$\varepsilon_{ij} = d_{kij} E_k + D_{ijkl} E_k E_l \, .$$

The deformations ε_{ij} in turn influence the optical properties:

$$\Delta a_{ij} = p_{ijkl} \varepsilon_{kl} \, .$$

d_{kij}, D_{ijkl} and p_{ijkl} designate the inverse piezoelectric, the quadratic electrostrictive and the elastooptical tensor. The relations between mechanically free and clamped coefficients are expressed by

$$r^\sigma_{ijk} - r^\varepsilon_{ijk} = p_{ijop} d_{kop} \quad \text{and}$$

$$R^\sigma_{ijkl} - R^\varepsilon_{ijkl} = p_{ijop} D_{opkl} \, .$$

2.5 Waveguides

The basic mechanism of a waveguide is easily demonstrated by the ray-optics approach. Consider an optical medium with three different regions as illustrated in Fig. 2.3. The necessary condition that a wave is confined in region II is $n_1 < n_2 > n_3$. The structure is assumed to be infinite in y-direction. At both interfaces total internal reflection occurs.

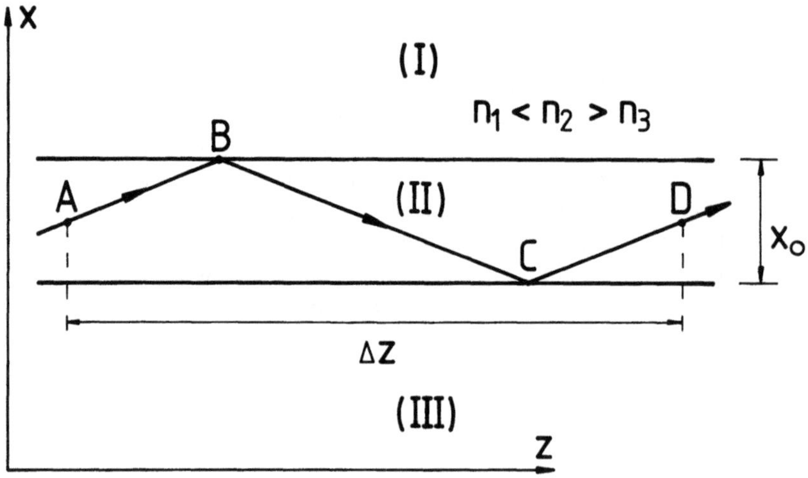

Fig. 2.3. Wave-guiding in a dielectric medium

The zigzagging plane wave sketched in Fig. 2.3 has the form

$$E(x, z, t) = E_0 \exp i(\omega t - \beta z - hx)$$

where β is the longitudinal and h the transversal wave number. At point B it experiences the phase shift $2\,\phi_{21}$ and at point C $2\,\phi_{23}$, both upon total reflection. ϕ_{21} and ϕ_{23} are related to the refractive indices by

$$\tan\phi_{21} = \sqrt{\frac{\beta^2 - n_1^2k_0^2}{n_2^2k_0^2 - \beta^2}} \quad \text{and} \quad \tan\phi_{23} = \sqrt{\frac{\beta^2 - n_3^2k_0^2}{n_2^2k_0^2 - \beta^2}}$$

for a y-polarization of $\vec{E}$ and by (2.8)

$$\tan\phi_{21} = \frac{n_2^2}{n_1^2}\sqrt{\frac{\beta^2 - n_1^2k_0^2}{n_2^2k_0^2 - \beta^2}} \quad \text{and} \quad \tan\phi_{23} = \frac{n_2^2}{n_3^2}\sqrt{\frac{\beta^2 - n_3^2k_0^2}{n_2^2k_0^2 - \beta^2}}$$

for a x-polarization of $\vec{E}$[8]. The total phase shift, which the wave experiences on its path from A to D is given by

$$\omega\Delta t - \beta\Delta z - 2\,hx_0 + 2\,\phi_{21} + 2\,\phi_{23}\,. \tag{2.9}$$

Not all rays which are reflected at the interface constitute a mode with energy transfer only in z-direction. A mode must have an unique propagation constant and a well-defined amplitude at each point in space and time[8]. It is described by

$$E = E(x)\exp[i(\omega t - \beta z)]\,. \tag{2.10}$$

Comparing the phase difference (2.9) with the phase in (2.10) we notice the restriction

$$-2\,hx_0 + 2\,\phi_{21} + 2\,\phi_{23} = 0 \tag{2.11}$$

or an integral multiple m of 2 π. Condition (2.11) differs for both polarization directions. The y-polarization of $\vec{E}$ is called the fundamental transversal electric (TE) mode and the x-polarization of $\vec{E}$, being at the same time a y-polarization of $\vec{H}$, is the fundamental transversal magnetic (TM) mode. Accordingly it is possible to distinguish between higher-order TE_m and TM_m modes, when (2.11) equals m2π.

2.6 Magnitude and Crystallographic Properties of Electrooptic Coefficients

As a general guide the magnitude of a linear electrooptic coefficient is 10^{-12} m/V and of a quadratic coefficient 10^{-21} m^2/V^2. Electrogyration effects are by the factor a/λ smaller than electrooptic effects, where a denotes a characteristic structural length and λ the wavelength of light. Actually r_{ij} may vary from $0 \pm 0.001 \times 10^{-12}$ m/V (r_{63} in $Sr_2ZnGe_2O_7$[9]) to 67×10^{-12} m/V (r_{11} in methyl-nitroaniline[10]) at room temperature. Similar large variations have been observed with the Kerr-effect. Usually the quadratic effects are much smaller than the linear ones. Considering a hypothetical crystal with the above mentioned values for r and R, with a refractive index of 1.5, an applied electric field of 10^4 V/cm, and a sample length of 1 cm, the induced phase differences for a wavelength of 550 nm are $2.1 \cdot 10^{-3}$ and $2.1 \cdot 10^{-8}$, respectively. The order of magnitude of electrogyration equals that of the Kerr-effect[11].

Applying an electric field, the electronic and ionic charge distributions are rearranged in a solid. The optical dielectric impermeability a_{ij} depends mainly on the valence electrons. A change in the ionic charge distribution, however, influences strongly the electronic potential and thereby the polarizability of the solid. Electronic resonances occur typically near 10^{15} Hz (300 nm) and lattice resonances in the neighbourhood of 10^{13} Hz. Thus pure electronic contributions to the electrooptic effect can be determined separately when the frequency of the modulation field is sufficiently high. Such experiments are carried out in nonlinear optics (SHG-effect, e.g.). At frequencies below the Tera-Hz region both contributions, the electronic r_{ij}^{e} and the ionic r_{ij}^{ion}, are measured. In the light of the bond charge model developed by Phillips and Van Vechten[12], the authors Levine[13] and Shih and Yariv[14] attribute r_{ij}^{e} to displacements of bond charges and r_{ij}^{ion} to changes of bond lengths and directions. Usually r_{ij}^{ion} exceeds r_{ij}^{e}. In KDP, for example, $r_{63}^{e} = 0.55 \times 10^{-12}$ m/V and $r_{63}^{ion} = 8.3 \times 10^{-12}$ m/V[15]. In CuCl the d-electrons enlarge r_{41}^{e}, but its sign is opposite to r_{41}^{ion}[16]. A possible way to increase electrooptical effects is the employment of frequencies near a resonance. For the optical region this means the use of small band gap materials. Excitations of lattice resonances is usually of no importance because in electrooptic devices the modulation frequency is much smaller than 10^{13} Hz. However, near a paraelectric-ferroelectric phase transition lattice resonance frequencies decrease drastically due to the softening of an optical phonon. In practice ferroelectric instability of a lattice is the most important phenomenon which establishes extraordinary large electrooptic constants. Thus most materials presented in Chap. 4 are ferroelectric crystals.

The above considerations are confined to physical properties, which can be brought to bear only if the structural situation enables the electrooptical effects. For large linear electrooptic constants the crystal structure must possess a high degree of acentricity or polarity, which is by no means a well defined quantity. The only sound method for its determination is the measurement of appropriate physical properties. In this connection the longitudinal piezoelectric effect and the generation of second harmonic light waves have to be mentioned. There exist a lot of empirical rules explaining the structural conditions for the existence of a large electrooptic constant. They have recently been elaborated by Bohatý[15] on the basis of an extensive comparison of experimental data. He found that favourable conditions exist, when a crystal belongs to a pyroelectric point group, contains strongly asymmetrical bonds and molecules with extreme hyperpolarizabilities or tends to be dielectrically unstable as in the case of ferroelectrics.

3 Electrooptic Modulation and Experimental Techniques

3.1 Induced Phase Retardation

In the expression

$$D_i = D_{0i} e^{i(\omega t - k x_3 + \phi)} \qquad (i = 1, 2)$$

for a plane electromagnetic wave, ϕ is considered as that part of phase retardation, the wave suffers on its path through a crystal plate:

$$\phi = \frac{2\pi}{\lambda_0} n_i L_3 .$$

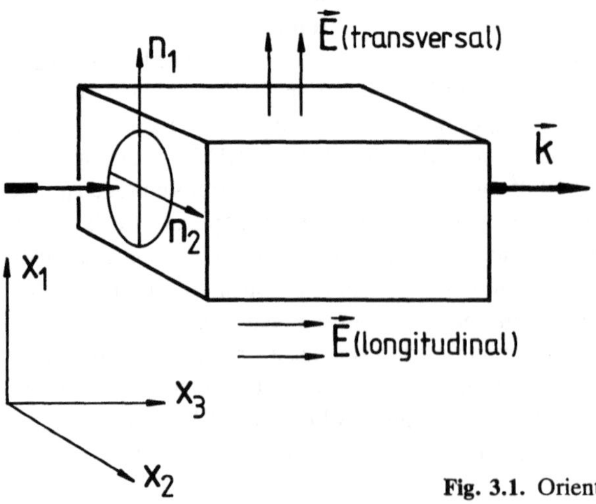

Fig. 3.1. Orientations in electrooptic measurements

Figure 3.1 shows the relative orientation of reference system and sample. Applying an electric field upon the sample, ϕ is changed accordingly to

$$\Delta\phi = \frac{2\pi}{\lambda_0}\left[L_3\Delta n_i + n_i\Delta L_3\right] . \tag{3.1}$$

From the expression $n_i = \sqrt{\frac{1}{a_i}}$ we obtain

$$\Delta n_i = -\frac{1}{2} n_i^3 \Delta a_i ,$$

where the increments Δa_i depend on the static electric field as shown by the equation

$$\Delta a_i = r_{ik}E_k + R_{ik}E_k^2 \qquad (i = 1 \text{ or } 2,\; k = 1 \text{ or } 3) .$$

Here r_{ik} and R_{ik} designate overall electrooptical coefficients, including all primary and secondary effects discussed in Chap. 2.4. The induced deformation $\varepsilon_3 = \frac{\Delta L_3}{L_3}$, the second term in (3.1), contains the electromechanical coupling

$$\varepsilon_3 = d_{k3}E_k + D_{3k}E^2 . \tag{3.2}$$

With this relation Eq. (3.1) reads in full length

$$\Delta\phi = \frac{2\pi}{\lambda_0}\left[-\frac{1}{2} n_i^3(r_{ik}E_k + R_{ik}E_k^2) + n_i(d_{k3}E_k + D_{3k}E_k^2)\right]L_3 . \tag{3.3}$$

Equations 3.1–3.3 are referred to the reference system which is shown in Fig. 3.1. x_i are the axes of a Cartesian coordinate system. The principal axes of the indicatrix are parallel to x_1 and x_2 and in the transversal configuration the electric field is along x_1 and in the longitudinal configuration $\vec{E}\|\vec{k}\|x_3$.

In all research papers, dealing with the determination of coefficients in important electrooptic materials, the induced changes of length have been neglected. Fortunately, in those crystals included in Chap. 4, the term $n_i d_{k3}$ is only a very small correction in the linear electrooptic response.

3.2 Interferometric Measurements

Figure 3.2 shows schematically a typical interferometric set-up. Other types of interferometers, like the Michelson-interferometer, are also in use. Let us assume that the electric field and the polarizations are along x_1. Then the phase difference between both beams is

$$\phi = \frac{2\pi}{\lambda_0}(n_1 - 1)L_3 ,$$

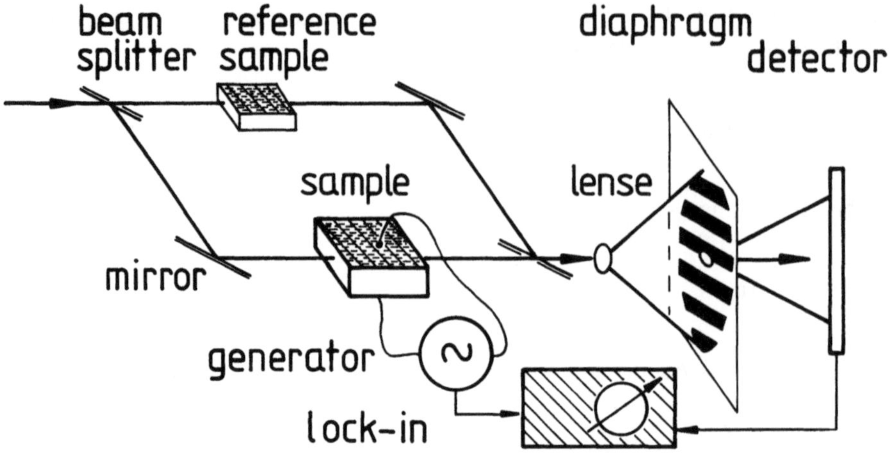

Fig. 3.2. Interferometric arrangement for the measurement of electrooptic coefficients

which varies with E_1 accordingly to

$$\Delta\phi = \frac{2\pi}{\lambda_0}\left[-\frac{1}{2}n_1^3 r_{11} + (n_1 - 1)d_{13}\right]L_3E_1 . \qquad (3.4)$$

Only linear effects have been taken into account. When the intensity in both arms of the interferometer is the same, the output intensity is given by

$$I = I_0\frac{1}{2}(1 + \cos\phi) = I_0\cos^2(\phi/2) .$$

Due to a small deviation of the propagation direction of both interfering waves the interference pattern, which is sketched in Fig. 3.2, can be obtained. Applying a dc-field, the fringes shift. Measuring this shift relative to the distance of two fringes, $\Delta\phi$ is determined. A more sensitive method is to select with a diaphragm a small part of the pattern and to observe the intensity changes

$$\Delta I = -I_0\sin\phi\cdot\Delta\phi .$$

In this case it is possible to apply alternating voltages and to detect ΔI with a lock-in-technique. Usually a sample with known electrooptical response is mounted in the second arm of the interferometer as a reference.

Interferometric measurements are most convenient for determining an electrooptic tensor as completely as possible. In addition the piezoelectric coefficients d_{ij} can be measured by using a sufficient amount of such orientations of the sample where different combinations of r_{ij} and d_{ij} are involved. In this way the electrooptic and the piezoelectric tensor of the orthorhombic crystal d(+)-camphoric acid acetone has been determined[17].

3.3 Induced Birefringence

In contrast to interferometric measurements the incoming wave is now polarized in a 45°-direction in respect of the principal axes of the indicatrix (Fig. 3.3). Due to the birefringence

$$\delta = \frac{2\pi}{\lambda_0}(n_1 - n_2)L_3$$

the emergent wave is elliptically polarized. As $D_{01} = D_{02}$ the principal axes of the polarization ellipse are also in a 45°-orientation. The following optical components are α) a λ/4-compensator and an analyzer or β) only an analyzer in the fixed position shown in Fig. 3.3. In the first case the λ/4-plate compensates the π/2-phase difference of both components of the elliptical wave. The resulting linear waves vibrates at an azimuthal angle

$$\eta = \frac{\delta}{2}\,.$$

Rotating the analyzer by an angle φ from the position, where the light is extinguished, the output intensity is

$$I = I_0 \sin^2\varphi\,.$$

I is changed by a modulating electric field accordingly to

$$\Delta I = -I_0 \sin 2\varphi \cdot \frac{1}{2}\Delta\delta \qquad (3.5)$$

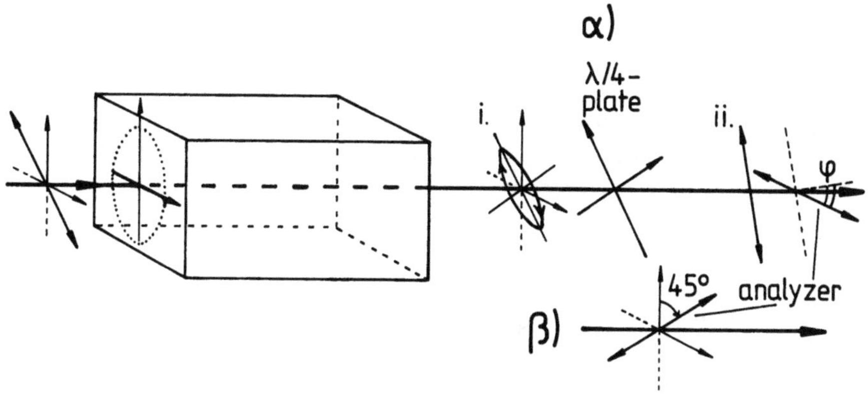

Fig. 3.3. Measurement of birefringence. The incoming linear wave vibrates 45° in respect to the principal axes of the indicatrix. The emergent wave *(i)* is elliptically polarized. In configuration α) the λ/4-compensator changes the elliptical wave into a linear one *(ii)* and the analyzer is rotated by an angle φ from the position of extinction. In configuration β) the positions of the polarizer and the analyzer are crossed

with

$$\Delta\delta = \frac{2\pi}{\lambda_0}\left[-\frac{1}{2}(n_1^3 r_{11} - n_2^3 r_{21}) + (n_1 - n_2)d_{13}\right]L_3E_1 \tag{3.6}$$

for the transversal configuration.

In deriving (3.5) we have considered only the term which is linear in $\Delta\delta$. In the description of the configuration β) of Fig. 3.3 we will include terms of higher order. The transmitted intensity in this arrangement is given by

$$I = I_0 \sin^2 \frac{\delta}{2} .$$

Expanding I in respect to δ yields

$$\Delta I/I_0 = \sin^2 \frac{\delta}{2} + \frac{1}{2}\sin\delta\,\Delta\delta + \frac{1}{4}\cos\delta(\Delta\delta)^2 - \frac{1}{12}\sin\delta(\Delta\delta)^3 + \dots$$

with $\Delta\delta$ of Eq. (3.6). Assuming a time dependent electric field

$$E = E_0 \sin \Omega t$$

and using an overall electrooptic constant r, the intensity varies with time as

$$I(t) = I_0\left[\frac{1}{2}\sin\delta \cdot rE_0 \sin\Omega t - \frac{1}{8}\cos\delta \cdot r^2E_0^2 \cos 2\,\Omega t - \right.$$

$$\left. \frac{1}{16}\sin\delta \cdot r^3E_0^3 \sin\Omega t + \frac{1}{48}\sin\delta \cdot r^3E_0^3 \sin 3\,\Omega t + \dots \right. \tag{3.7}$$

The different terms in (3.7) can be elaborated by observing I(t) at different frequencies as a function of field strength and by adjusting δ to $m \cdot \pi/2$ or $m \cdot \pi$, where m is an integral number. In this way it is also possible to separate the linear effect from the quadratic effect, which is not considered in (3.7). Making allowance for the Kerr-effect, I(t) at frequency 2 Ω reads

$$I(2\,\Omega) = -I_0\left[\frac{1}{2}\sin\delta \cdot R + \frac{1}{8}\cos\delta \cdot r^2\right]E_0^2 \cdot \cos 2\,\Omega t .$$

By a variation of δ, which is achieved easily with an additionally applied dc-voltage, R and r can be determined.

The techniques described in Sect. 3.2 and 3.3 as methods for the determination of electrooptic coefficients are generally the same as are used in devices for the modulation of intensities.

3.4 Induced Birefringence in the Presence of Optical Activity

Usually optical activity is much smaller than natural birefringence and therefore it can be neglected in electrooptical measurements. However, in cubic crystals with symmetry 23 and in special birefringent crystals with the optical beam parallel to an optical axis, the induced birefringence $\Delta\delta$ interacts with the specific rotation ϱ. Measurements with electric fields at zero frequency have been reported by Sliker[18]. Using alternating fields Weber[19] obtained for the amplitude of modulation

$$\Delta I = I_0 \frac{\sin \varrho L}{\varrho L} \sin(2\chi + \varrho L) \sin 2\varphi \cdot \frac{1}{2} \Delta\delta . \tag{3.8}$$

This expression is valid for the experimental set-up shown in Fig. 3.4. Equation (3.8) has been derived by assuming $2\varrho \gg \delta$ which holds for a light propagation nearly parallel to an optical axis and in an isotropic medium. For a nongyrotropic medium the expression equivalent to (3.8) reads

$$\Delta I = I_0 \sin 2\chi \cdot \sin 2\varphi \cdot \frac{1}{2} \Delta\delta .$$

The main differences in the study of both kinds of materials is the different orientation of the polarizer and the compensator (Fig. 3.4) and the term $\sin \varrho L/\varrho L$, which always

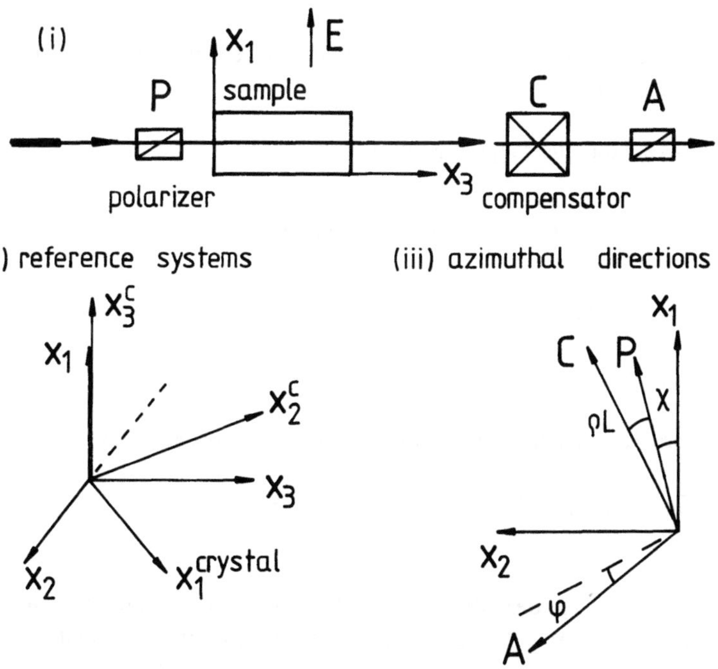

Fig. 3.4. Experimental configuration for measuring the electrooptical effect in an optically active crystal of symmetry 23. ϱL is the natural rotation of a sample with the length L. φ is the deviation of the position of the analyzer *(A)* from extinction

reduces the modulation efficiency in a gyrotropic crystal. In respect to the orientation of the polarizer intensity modulation is maximal for $\chi = 45° - \varrho L/2$ (gyrotropic) and $\chi = 45°$ (non-gyrotropic), respectively.

It is also possible to carry out the electrooptic investigation with circularly polarized waves by mounting the compensator in Fig. 3.4 between the polarizer and the sample. With this method Pellat-Finet[20)] determined recently the coefficient r_{41} of $Bi_{12}SiO_{20}$ which is an important electrooptical crystal possessing optical activity.

3.5 Frequency Modulation

In Sect. 3.3 we have already shown that with an alternating electric field of frequency Ω signals at Ω and $m \cdot \Omega$ occur, where m is an integral number. The equations given are valid only for low modulation frequencies. If Ω is no longer negligible compared to the optical frequency ω, a different treatment of the problem becomes necessary. Assuming again a modulating field of the kind

$$E = E_0 \sin \Omega t$$

and making use of only one optical mode, the light wave is expressed by

$$D = D_0 \exp i \frac{2\pi}{\lambda_0} \left[n - \frac{1}{2} n^3 \cdot r \cdot E_0 \cdot \sin \Omega t - ct \right] .$$

n and r are the actual refractive index and electrooptic coefficient. Omitting constant phase factors,

$$D = D_0 \exp(-i\omega t) \exp(i\Delta\phi \sin \omega t) . \tag{3.9}$$

$\Delta\phi = -\frac{\pi}{\lambda_0} n^3 r E_0$ is the modulation index.

Using the identity

$$\exp i(x \sin \theta) = \sum_{m=-\infty}^{\infty} J_m(x) e^{im\theta} ,$$

where $J_m(x)$ designates the Bessel-function of order m, Eq. (3.9) can be rewritten as

$$D/D_0 = J_0(\Delta\phi)e^{-i\omega t} + J_{-1}(\Delta\phi)e^{-i(\omega+\Omega)t} + J_1(\Delta\phi)e^{-i(\omega-\Omega)t} + J_{-2}(\Delta\phi)e^{-i(\omega+2\Omega)t} + J_2(\Delta\phi)e^{-i(\omega-2\Omega)t} + \ldots \tag{3.10}$$

Equation (3.10) describes the distribution of energy in the fundamental wave and in the optical sidebands. The Bessel functions depend on the modulation index $\Delta\phi$:

$$J_0(\Delta\phi) = 1 - \left(\frac{\Delta\phi}{2}\right)^2 + \frac{1}{4}\left(\frac{\Delta\phi}{2}\right)^4 - \frac{1}{36}\left(\frac{\Delta\phi}{2}\right)^6 + - \ldots$$

$$J_{-1}(\Delta\phi) = \frac{\Delta\phi}{2} - \frac{1}{2}\left(\frac{\Delta\phi}{2}\right)^3 + \frac{1}{12}\left(\frac{\Delta\phi}{2}\right)^5 - + \ldots$$

$$J_{-2}(\Delta\phi) = J_2(\Delta\phi) = \frac{1}{2}\left(\frac{\Delta\phi}{2}\right)^2 - \frac{1}{6}\left(\frac{\Delta\phi}{2}\right)^4 + \frac{1}{48}\left(\frac{\Delta\phi}{2}\right)^6 - + \ldots \quad (3.11)$$

Inserting (3.11) into (3.10) we obtain for the quotient of the second over the first sideband approximately $\Delta\phi$. Therefore the electrooptic coefficients can be determined by measuring the relative intensities of the sidebands.

As an example, let us use as light source a He-Ne-laser of $\lambda = 632.8$ nm and a modulation frequency of $\nu_m = \Omega/2\pi = 10^7$ Hz. Then the frequency shift $\nu - \nu_m$ corresponds to a wavelength shift of $\Delta\lambda = 0.013$ nm. This is the same magnitude occuring in Brillouin scattering and can be resolved with a Fabry-Perot-etalon.

Kaminow[21)] modulated a He-Ne-laser beam at 9.01 Gc/s with a KDP crystal and observed the sidebands photographically as a splitting of the rings produced by a Fabry-Perot interferometer. Employing a far-infrared laser to drive a modulation wave, even higher frequencies are obtainable. With a cw HCN laser the electrooptic coefficients of $LiNbO_3$ could be measured at 964 GHz and at 337 GHz[22)]. Within experimental uncertainty the results agreed with values obtained at the radio-frequency of 65 MHz. This invariance of the electrooptic effect in $LiNbO_3$ is not surprising because the lowest resonance frequency of optical phonons occurs at 7.5 THz, well above 964 GHz[22)]. At frequencies essentially higher than 1 GHz, the electric field is not instantaneous within the whole crystal volume but acts as a traveling wave and special care has to be taken to match the phase velocities of the interacting waves.

Another successful method for high frequency measurements is the optical heterodyne technique, which has been employed for the first time by Kaminow on $BaTiO_3$[23)]. A critical test of this method was performed by de Lange[24)] with an experimental set-up shown schematically in Fig. 3.5. The frequency of the local-oscillator laser was kept 70 MHz apart from the signal laser by means of an automatic-frequency-control

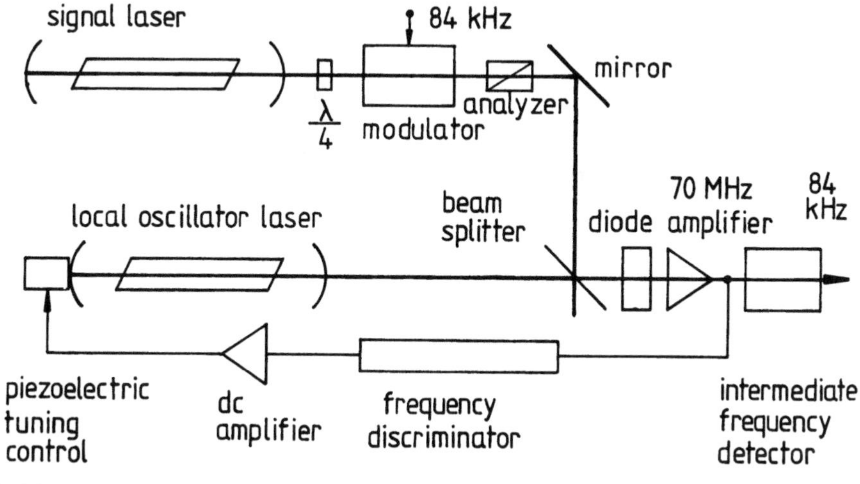

Fig. 3.5. Electrooptic measurements by the heterodyne technique

circuit. Both laser beams were combined at the beam splitter, yielding a carrier frequency at 70 MHz for the modulation signal at 84 kHz. In the experiment of Kaminow[23] the modulation frequency ν_m (30 to 70 MHz) was of the same magnitude as the intermediate frequency ν_i of 70 MHz. Thus signals at ν_m and ν_i together with the corresponding sidebands were observed. In comparison with a direct detection the heterodyne receiver appears to be more satisfactory for the reception of phase- or frequency-modulated light and allows a transformation of optical frequencies, which may be advantageous at some wavelengths in respect of detection and amplification[24].

3.6 Pulse Technique

As described in the above section, high-frequency techniques require large experimental efforts. Comparatively more simple measurements of the clamped electrooptic coefficients can be carried out by the pulse method. A typical transient optical response of the crystal to an applied voltage step is illustrated in Fig. 3.6. The time dependence is observed by feeding the photomultiplier output directly into a sampling oscilloscope[25] or by boxcar integration[26]. In Fig. 3.6 the first horizontal plateau measures the clamped electrooptical response. After a delay of 3 μsec the piezoelectrically excited acoustic waves deform the sample and an additional strain-optical contribution is observed. In detail the optical response may look rather complicated due to the interference of different acoustic waves[26]. The pulse-method has been used for the determination of electrooptic coefficients in $BaTiO_3$[25], in GaP[27], and in KDP, DKDP and $LiCOOH \cdot H_2O$[26].

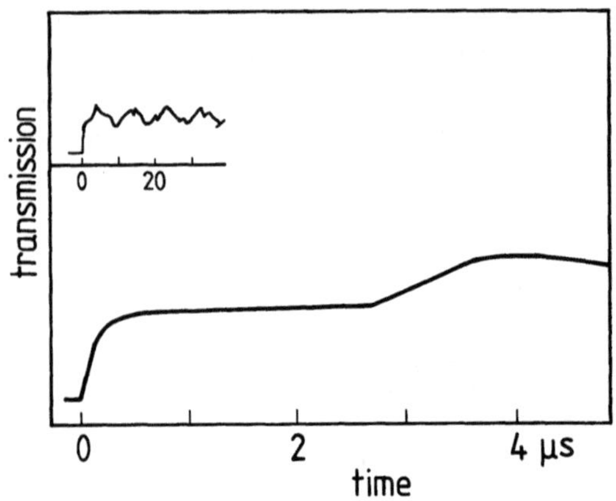

Fig. 3.6. Electrooptic response to a rectangular pulse of 600 μs in KDP[26]. Only the beginning of the pulse response is shown

4 Electrooptic Crystals

4.1 Selection of Materials

A complete survey about electrooptic materials and their coefficients is available in two editions of Landolt-Börnstein[28]. Additional information has recently been reported by Bohatý[15]. Here we concentrate on such crystals which are important in respect of applications. From a crystal-chemical standpoint three different groups may be distinguished:

A) Crystals with oxygen-octahedra. To this class belong ABO_3-perovskites, tetragonal bronzes and $LiNbO_3$.

B) Ionic crystals containing molecular groups. The most important members are KDP and its isotypes.

C) Semiconductors.

In the above survey two classes of materials are missing. Since Kurtz and Perry[29] developed a powder technique for the evaluation of nonlinear optical materials it has become known that intense generation of second harmonic of laser radiation can be achieved with special molecular organic crystals.

Charge transfer transitions are the origin of their large effects[30]. Such a material, 2-methyl-4-nitroaniline (MNA), has been suggested as an effective electrooptical crystal[31]. It is also a nice example for the potentiality of controlled chemical engineering. Starting from a benzene ring, a NH_2 and a NO_2 group are substituted in directly opposite positions. NH_2 serves as a donor and NO_2 as an acceptor. Furthermore a methyl group affects a noncentrosymmetric crystal structure[31]. The electrooptic coefficient $(n_1^3 r_{11} - n_3^3 r_{31})$ has been determined as $540 \pm 100 \times 10^{-12}$ m/V. This is more than five times the induced birefringence in $LiNbO_3$. But only small crystals can be grown and inhomogeneities can reduce the electrooptic figure of merit by more than one order of magnitude[32]. The second kind of materials which is out of the scope of the present study are ferroelectric ceramics. Their main components are the perovskite oxydes $BaTiO_3$, $SrTiO_3$, $PbTiO_3$ and $PbZrO_3$[33]. Especially solid solutions of $PbTiO_3$ and $PbZrO_3$ (PZT), doped with La (PLZT-ceramics) are used for practical applications. Ferroelectric ceramics have the advantage of easy preparation and their composition can be varied in a wide range to adjust the properties of the material. As the grains are randomly orientated, the macroscopic effects are usually smaller than in a single crystal with the same composition. Grain boundaries may cause distortions of the light wave, especially in the presence of electric fields. However, in the ceramic composition $Pb_{0,90}La_{0,10}(Zr_{0,65}Ti_{0,35})O_3$ this influence on phase modulation has been identified as less important compared to interference effects by surface reflection and nonuniformity of the electric field[34].

The prevailing impression one gets from the publications about devices of recent years, is the dominance of $LiNbO_3$ in respect of materials used. Therefore this crystal is described more thoroughly in the next section. The structure of tetragonal bronzes, described in Sect. 4.3 allows a high variation of chemical constituents and thereby a tailoring of effective materials. Some of the early devices had been fabricated from perovskites and many fundamental concepts were investigated with $BaTiO_3$. The outstanding feature of these crystals is their paraelectric-ferroelectric phase transition. The influence of such a transition on electrooptic properties will be demonstrated in Sect. 4.4 with the example of $BaTiO_3$. Although the refractive indices of KDP-type compounds are comparatively small, they are the basis of many devices. The reason is the easy availability of large single crystals with high optical quality. Therefore a lower modulation efficiency can be compensated by employing large samples. KDP is the second example, with which the determination of coefficients is illustrated in more detail in Sect. 4.5. Semiconductors are without doubt the most important non-metallic compounds in technical applications. The feasibility of electrooptic semiconductor devices, however, is restricted by small coefficients and problems in applying high electric fields. They are described in the last section of this chapter together with two-dimensional heterostructures. In the last few years much interest has been focused on the latter materials because they allow an integration of light sources, optical modulators, and detectors in one optical circuit.

The magnitude of coefficients is not the only criterion for applications. Of the same importance are crystal growth properties, refractive indices, transparency and dielectric properties. Some of these aspects are considered in this chapter.

4.2 $LiNbO_3$ and $LiTaO_3$

4.2.1 Crystal Growth

At room temperature $LiNbO_3$ is a ferroelectric crystal with the rhombohedral space group R3c. The transition to the paraelectric phase (space group $R\bar{3}c$) occurs at 1210 °C, which is just below the melting temperature of 1253 °C[35]. $LiNbO_3$ melts congruently and can easily be grown as large single crystals by the Czochralski technique[36]. The starting material is prepared by mixing and sintering Nb_2O_5 and Li_2CO_3 in stoichiometric amount. The most suitable crucible material is platinum[37]. Due to oxygen deficiency and partial reduction of niobium, as-grown crystals show a pale brown colour. They are bleached by annealing at 1100 °C in air or oxygen-atmosphere. Different kinds of defects have been detected in Czochralski pulled lithium niobate: solid phase inclusions, gaseous inclusions, low angle grain boundaries, twin planes, and dislocations[37]. Most of these defects can be avoided by carefully observed growth conditions. Under these the selection of seeds is important. They should be free from twin boundaries and should possess an optimal crystallographic direction. Furthermore 180°-domains occur due to the paraelectric-ferroelectric phase transition. They can be avoided by adding 0.5 at.% MoO_3 to the melt or by performing the growth in an electric field. It is also possible to homogenize a sample by electrical reversal of the spontaneous polarization at about 1200 °C[37]. Some properties of $LiNbO_3$ can be changed in a controlled way by doping the material with MgO[38].

Crystallographic properties and crystal growth of $LiTaO_3$ are similar to those of $LiNbO_3$. The melting temperature is higher (1650 °C) and the phase transition lower (665 °C) in $LiTaO_3$[35].

Both materials, $LiNbO_3$ and $LiTaO_3$ are used as waveguides, which can be produced by different techniques. For instance, by liquid phase-epitaxy in a suitable flux a Li-rich film is grown on $LiNbO_3$- or $LiTaO_3$-samples. $LiNbO_3$, for example, can be crystallized in the nonstoichiometric form $(Li_2O)_v(Nb_2O_3)_{1-v}$ with an index variation of $\partial n_e/\partial v = -1.63$[39]. Nowadays the favoured method for fabricating $LiNbO_3$-waveguides is indiffusion of titanium. Strips of Ti-layers of 20–80 nm thickness are evaporated onto a polished crystal face[40]. The orientation depends on the planned application and is usually (100) or

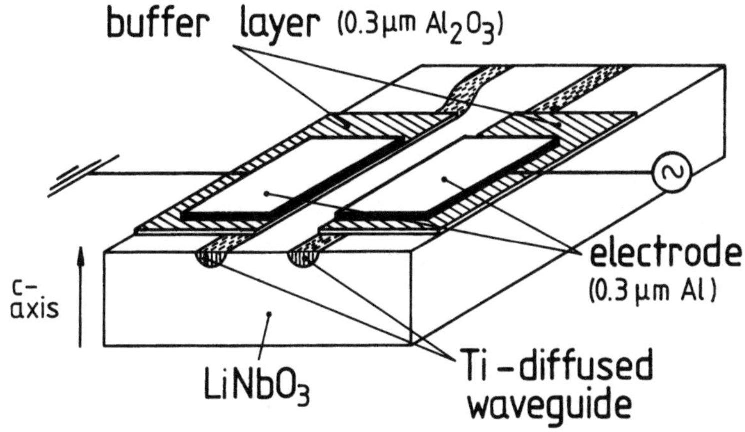

Fig. 4.1. Structure of a $LiNbO_3$-waveguide modulator. The electrodes are placed in such a way that the main field direction is parallel to the crystallographic c-axis for utilizing the large electrooptic coefficient r_{33}. After Ref. 42

(001). The sample is held for several hours at 850–1000 °C. To prevent oxydation of the metal Ar is passed over it. At the end of the diffusion process Ar is replaced by oxygen to reoxidize $LiNbO_3$. The important waveguide parameters – number of modes, change of refractive index, and effective guide thickness – can be independently controlled by the diffusion parameters: thickness of metal layer, temperature and diffusion time[40)]. Outdiffusion of Li_2O, leading to undesired surface guiding, is eliminated by using wet Ar-gas[41)]. A typical $LiNbO_3$-waveguide is exhibited in Fig. 4.1[42)]. The function of the buffer layer is to avoid absorption loss by the metal electrode and to obtain a uniform electric field distribution in the waveguide.

4.2.2 Measurements of Free Coefficients

In point group 3 m four independent linear electrooptic coefficients exist. The r_{ij}-matrix is shown in Table 4.1. According to the recommendations of IRE-standards[43)] for piezoelectric crystals the physical reference system x_i is chosen in the following way: x_1 is perpendicular to the mirror plane and the positive end of x_3-axis becomes electrically negative when the crystal is compressed along x_3, which coincides with the crystallo-

Table 4.1. Matrix of electrooptic coefficients r_{ij} in point group 3 m

0	$-r_{22}$	r_{13}
0	r_{22}	r_{13}
0	0	r_{33}
0	r_{51}	0
r_{51}	0	0
$-r_{22}$	0	0

Table 4.2. Configurations for determining the electrooptic coefficients in $LiNbO_3$

	$\vec{k}$	$\vec{D}$	$\vec{E}$	Coefficients[a]	Numerical values[b]	Ref.	$2(n' - n'')d/n^3$	Corrected values[b]
A1	x_1	x_3	x_3	$r_{33} - 2(n_e - 1)d_{31}/n_e^3$	30.9	45, 44	−0.2	30.7
					34.0	46		33.8
A2	x_1	x_2	x_3	$r_{13} - 2(n_0 - 1)d_{31}/n_0^3$	9.6	45, 44	−0.2	9.4
					10.9	46		10.7
A3	x_1	x_2	x_2	$r_{22} + 2(n_0 - 1)d_{22}/n_0^3$	6.81	45	4.5	2.3
A4	$x_1 + x_3$	$x_3 - x_1$	$x_3 - x_1$	$\frac{1}{4}\sqrt{2}\,[r_{13} + r_{33} + 2\,r_{51} - 2(n_e' - 1) \cdot$ $\cdot\,(d_{31} + d_{33} - d_{15})/n_e'^3]$[c]				
B1	x_1	$x_2 + x_3$	x_3	$(n_e^3/n_0^3)r_{33} - r_{13} - 2(n_e - n_0)d_{31}/n_0^3$	17.4[d]	47	0.01	17.4
					18.0	48		18.0
B2	x_1	$x_2 + x_3$	x_2	$-r_{22} + 2(n_e - n_0)d_{22}/n_0^3$	−3.3[e]	47	−0.3	−3.0
					−7.33	48		−7.03
B3	x_2	$x_1 + x_3$	x_2	$r_{22} - 2(n_e - n_0)d_{22}/n_0^3$	3.3[e]	47	−0.3	3.0
					6.0			5.7
B4	x_2	$x_1 + x_3$	x_3	$(n_e^3/n_0^3)r_{33} - r_{13} - 2(n_e - n_0)d_{31}/n_0^3$	17.4[d]	47	0.01	17.4
					17.0	48		17.0
B5	x_3		x_1	r_{22}	6.7	48	–	6.7
B6	x_3		x_2	r_{22}	6.3	48	–	6.3

[a] The signs are reversed compared with Eq. (3.4) and (3.6); [b] Unit: 10^{-12} m/V; [c] $\frac{1}{n_e'^2} = \frac{1}{2}\left(\frac{1}{n_e^2}\right) + \frac{1}{n_0^2}$; [d, e] The correct orientations are not known

graphic c-axis. Observing this convention, Hulme et al.[44] determined the signs of all four coefficients to be positive. Four different methods have been reported for the determination of mechanically free coefficients in $LiNbO_3$:

A. Zook et al.[45] have measured induced phase retardations $\Delta\phi$ with a Michelson-interferometer. The relation between $\Delta\phi$ and the electric field has been given in Eq. (3.4). The realized configurations of wavevector $\vec{k}$, polarization direction $\vec{D}$, and electric field direction $\vec{E}$ are quoted in Table 4.2 (No. A1–A3) together with the involved coefficients. Similar measurements with a modified Mach-Zehnder-interferometer have been performed by Onuki et al.[46]. They used the orientations A1 and A2 in Table 4.2. Configuration A4 has been added as a possible example for determining the tensor constant r_{51}, but it has never been applied with $LiNbO_3$.

B. With the method described in Chap. 3.3 Bernal et al.[47] have determined $r_c = \left(\frac{n_e}{n_0}\right)^3 r_{33} - r_{13}$ and r_{22}, which corresponds to B1 and B2 in Table 4.2. The only authors who observed the same coefficients in different directions and tested the tensorial behaviour of the effect, are Lenzo, Spenger and Nassau[48] (No. B1–B6 in Table 4.2).

In addition to measurements A and B which are listed in Table 4.2 the following experiments have been reported:

C. Bernal et al.[47] have determined r_{51} by observing the rotation of the optical indicatrix when the electric field is applied along x_1. They obtained $r_{51} = 32 \times 10^{-12}$ m/V.

D. Iwasaki et al.[49] observed conoscopically the field induced angle of optical axes V:

$$\tan V = \sqrt{n_0^3 \frac{r_{22}E_2}{n_e - n_0}}$$

in a (001)-plate. They determined $r_{22} = -6.4 \times 10^{-12}$ m/V. Although they orientated the sample according to I.R.E.-standards, the sign disagrees with that observed by Hulme et al.[44].

All authors cited under method A and B have neglected the induced change of length. This fault requires a critical examination of the published data.

As shown in Table 4.2 in the measurements B5 and B6 of Lenzo et al.[48] no piezoelectric change of sample length is involved. Thus we conclude that the difference of both values reflects the experimental accuracy. Then the true value is $r_{22} = 6.5 \pm 0.2 \times 10^{-12}$ m/V. The contribution $(n' - n'')d$ has to be subtracted from all other values. Piezoelectric constants and refractive indices used for these calculations are gathered in Table 4.3. The corrections employed do not change the original values significantly. Nevertheless, it is interesting to note, that they reduce the discrepancies observed by

Table 4.3. Piezoelectric (d_{ij}) and elastooptic coefficients (p_{ij}) and refractive indices of $LiNbO_3$ for $\lambda = 633$ nm

$d_{31} = -0.91$	$p_{11} = -0.0023$	$p_{31} = 0.167$
$d_{33} = 6.0$	$p_{12} = 0.081$	$p_{33} = 0.077$
$d_{22} = 20.9$	$p_{13} = 0.111$	$p_{41} = -0.138$
$d_{15} = 70.4$	$p_{14} = -0.073$	$p_{44} = 0.161$
unit: 10^{-12} C/N	$n_e = 2.2003$	$n_0 = 2.2873$

Table 4.4. Mechanically free electrooptic coefficients of $LiNbO_3$; $\lambda = 633$ nm; unit 10^{-12} m/V

$r_{22} = 6.4$	$r_{13} = 10.4$
$r_{33} = 31.8$	$r_{51} = 32$
$r_c = \frac{n_e^3}{n_0^3} r_{33} - r_{13} = 17.9$	

Lenzo et al.[48]. This fact makes the positive sign of r_{22} the most likely one. Both measurements of $r_c = \left(\frac{n_e}{n_0}\right)^3 r_{33} - r_{13}$ are practically the same. Thus their average 17.7×10^{-12} m/V is assumed to be the correct value. Comparing this coefficient with the differences A1–A2, the results of Ref. 45 ($r_c = 17.9 \times 10^{-12}$ m/V) seem to be more reliable than those of Ref. 46 ($r_c = 19.4 \times 10^{-12}$ m/V). A3 and B2b in Table 4.2 seem to be wrong. From preceding considerations we suggest for $LiNbO_3$ the electrooptical constants displayed in Table 4.4.

4.2.3 Mechanically Clamped Coefficients

The problems arising from the change of sample length are absent in measurements of the clamped coefficients r_{ij}^{ε}. Experiments at high frequencies have been carried out by Turner[50] and Kaminov et al.[51]. Their results are presented in Table 4.5. Oviously the experimental accuracy is poorer than in low-frequency-measurements. Therefore additionally the more accurate r_{ij}^{σ} are used to calculate r_{ij}^{ε} by the relation

$$r_{ij}^{\varepsilon} = r_{ij}^{\sigma} - \sum_k p_{ik} d_{jk} .$$

Table 4.5. Mechanically clamped electrooptic coefficients of $LiNbO_3$; $\lambda = 633$ nm; unit 10^{-12} m/V

	r_{22}	r_{13}	r_{33}	r_{51}
a	3.4	8.6	30.8	28
b	3.4	7.7	28.8	18.2
c	−6.8	0.6	0.16	5.6
c′	3.4	0.6	0.16	17.1
d	13.2	9.8	31.6	26.4
d′	3.0	9.8	31.6	14.9
e	3.3	8.7	30.4	16

[a] Turner[50]; [b] Kaminow et al.[51]; [c, c′] strainoptical contribution $p_{ik}d_{jk}$; [d, d′] $r_{ij} - p_{ik}d_{jk}$; unprimed notation [c, d]: all signs of p_{ij} positive; primed notation [c′, d′]: p_{11}, p_{14} and p_{41} negative; [e] averaged values

The summation of several coefficients makes the calculations convenient only if p_{ij} and d_{ij} are known very exactly. This is the case for the piezoelectric coefficients. The values in Table 4.3 have been averaged from three different authors[28)]. Discrepancies do not exceed 5%. The situation is less promising for the elastooptic coefficients, because the results of different authors vary widely[35)]. We have averaged the magnitudes of reported values (Table 4.3). Whereas most authors notify no sign, Avakyants[52)] suggests negative coefficients p_{11}, p_{14} and p_{41}. This is curious, especially for p_{11}, where no shear strain and no rotation of the optical indicatrix is involved. A negative p_{11} means an increase of optical polarizability, in spite of a decrease of density, for elongation along x_1. Both choices of sign are comprehended in Table 4.5. The good agreement of $r^{\varepsilon}_{22} = 3.0 \times 10^{-12}$ m/V with the high frequency values suggests the validity of Avakyants' signs. r^{ε}_{13} and r^{ε}_{33} agree in both kinds of experiments and $r^{\varepsilon}_{51} = 28 \times 10^{-12}$ m/V seems to be erroneous.

Electrooptic coefficients of $LiTaO_3$ are similar to those of $LiNbO_3$.

4.3 Tetragonal Bronzes

4.3.1 Characterization of Crystals

The general chemical formula of tetragonal bronzes is $A_{\alpha}B_{\beta}C_{\gamma}R_{10}O_{30}$, where A, B, and C refer to different cation sites in the unit cell. RO_6-octahedrons form a two-dimensional network perpendicular to the short c-axis, which has the length of a RO_6-diagonal, about 3.9 Å. In Fig. 4.2 three different RO_6-rings are distinguished, built up by 4, 5, and 3 octahedrons. Ten gaps of three different sizes can be occupied by the cations A, B, and

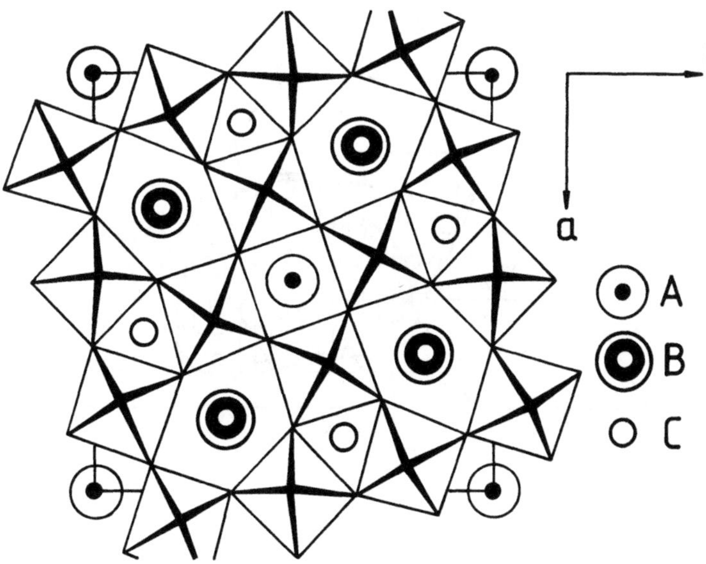

Fig. 4.2. Crystal structure of tetragonal bronzes. In $Ba_2NaNb_5O_{15}$ *A* and *B* are sodium and barium ions; *C*-sites are empty

C. 2 gaps A are surrounded by 12, 4 gaps B by 15, and 4 gaps C by 9 oxygen-ions. The kind of gap, a cation occupies depends mainly on its size. A and B are cations of larger and medium size as Na, K, Ca, Sr, Ba, Pb, SE. Li is a typical medium size cation C. Representatives of the sixfold coordinated cation R are Nb, Ta, and W[54)].

The most prominent member of this family is $Ba_2NaNb_5O_{15}$, abbreviated Banana or BNN. In BNN the sizes C are empty and the Na- and Ba-ions are distributed on gaps A and B. This distribution is not unique, it depends on chemical composition and crystal growth conditions. In $Ba_{4.13}Na_{1.74}Nb_{10}O_{30}$, for instance, position A is occupied by 87% Na and 6.5% Ba[55)]. The kind of distribution influences physical properties such as ferroelectricity significantly. In the ternary system Na_2O-BaO-Nb_2O_5 (Fig. 4.3) the liquidus isotherms demonstrate the relative flat liquidus near $NaBa_2Nb_5O_{15}$. The binary join $NaNbO_3$-$BaNb_2O_6$ reveals a steep solidus. Little solid Na_2O-solubility, flat liquidus, and low viscosity suggest crystal growth on the Na_2O-rich part of the diagram[56)]. According to Fig. 4.4 growth of selected solid compositions is possible by varying the melt composition from 60 to 80 mol% $BaNb_2O_6$. With increasing alkaline earth concentration the paraelectric-ferroelectric transition temperature T_c decreases. In pure $Ba_2NaNb_5O_{15}$ $T_c = 560\,°C$ and in crystals of 80 mol% $BaNb_2O_6$ (the upper bronze (ss) boundary in Fig. 4.4) $T_c = 509\,°C$[57)].

Large single crystals of different composition have been obtained employing the Czochralzki-method under conditions, which are typical for oxygen containing materials[58)]. As-grown crystals are usually coloured blue due to a Nb^{4+} oxydation state. Two distinct annealing effects have been observed. The first is a bleaching of crystals and the

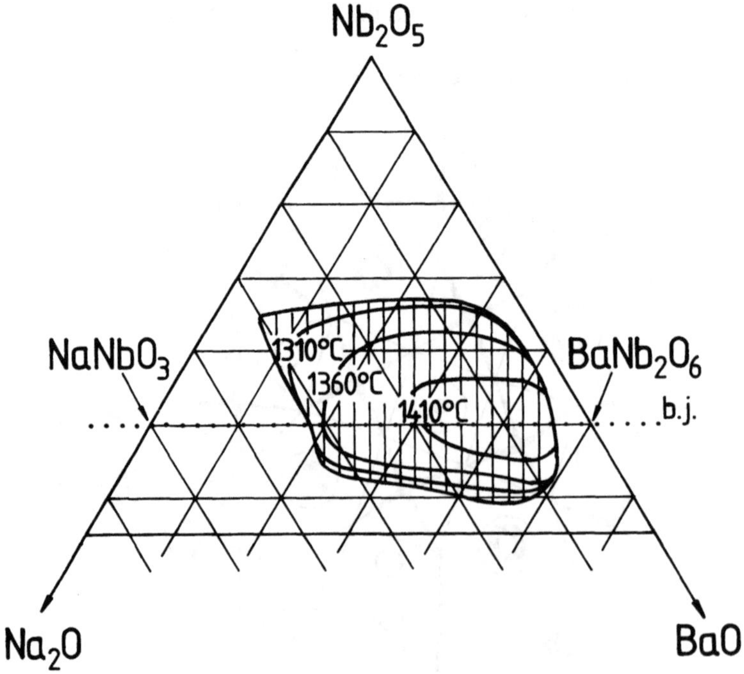

Fig. 4.3. Niobium rich part of the ternary phase diagram Na_2O-Nb_2O_5-BaO[56)]. The hatched region represents the compositions of the melt, from which growth of a bronze solid is possible. The binary join b.j. is shown in the next figure

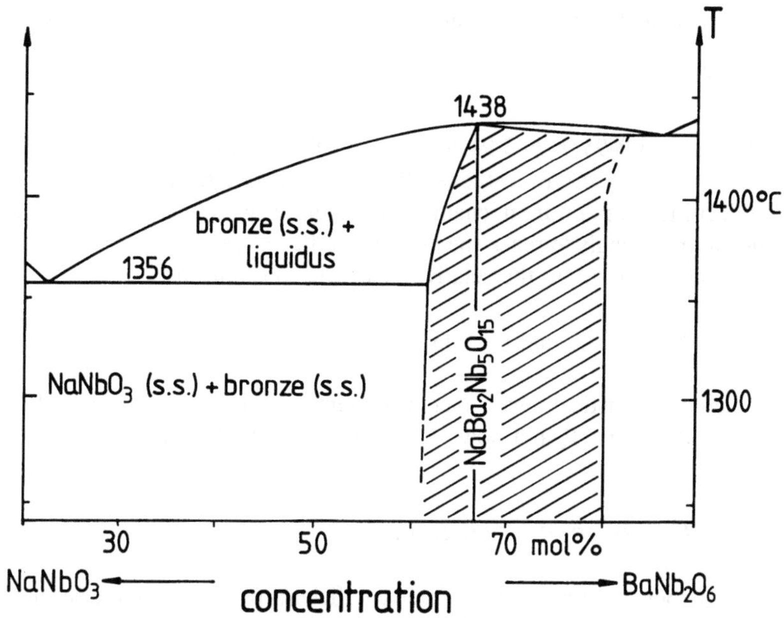

Fig. 4.4. Section of the phase diagram $NaNbO_3$-$BaNb_2O_6$[57]. The hatched region represents the existence of tetragonal bronze solid solutions

second involves shifts of the paraelectric-ferroelectric phase transition. Burns and O'Kane[59] have proved that the T_c of a pure BNN-crystal depends on the temperature from which it is quenched. A comparison of the quenching effect in $Ba_2NaNb_5O_{15}$, $Ba_{2.22}Na_{0.56}Nb_5O_{15}$, and $Ba_2K_xNa_{1-x}Nb_5O_{15}$ identified very clearly the structural disorder of Na- and Ba-cations as the origin of T_c-decrease.

A lot of other ferroelectric bronzes exist with even lower phase transition temperatures than in BNN. In some types the structural disorder is so high that diffuse transitions are observed[60].

At 300 °C BNN suffers a second structural phase transition from the tetragonal space group P4bm to the orthorhombic space group Ccm2. The orthorhombic x_1-x_2-plane is rotated through nearly 45° in respect to the tetragonal x_1-x_2-axes, whereas both x_3-axes remain parallel. The 300 °C-transition easily causes 45°-twins. Single domain samples are obtained by heating the crystal to a temperature above 300 °C and cooling it down under a compressive stress perpendicular to [001][61]. Physical properties are not changed significantly by the tetragonal to orthorhombic phase transition. Most of the optical, dielectric, and electrooptic properties described below have been reported by Singh, Draegert, and Geusic[61].

4.3.2 Optical and Dielectric Properties

BNN is fundamentally transparent from the band edge near 370 nm to 5 μm. Usually a narrow band near 3 μm occurs, which has been assigned to OH-stretching vibrations.

The dispersion of refractive indices fit the single-term Sellmeier equation

$$n_i^2 - 1 = s_i\lambda^2/(\lambda^2 - \lambda_i^2)$$

where $s_1 = 3.9495$, $s_2 = 3.9495$, $s_3 = 3.6008$, $\lambda_1 = 200.97$ nm, $\lambda_2 = 200.35$ nm, and $\lambda_3 =$ 179.44 nm. These data establish the practical tetragonal symmetry of the compound. With rising temperature n_2 changes only slightly and n_3 increases up to 300 °C linearly and then more strongly to reach the n_2-value at the ferroelectric-paraelectric phase transition. At room temperature the relative dielectric constants are

$$\bar{\varepsilon}_1 = 238,\ \bar{\varepsilon}_2 = 228,\ \text{and}\ \bar{\varepsilon}_3 = 43\ .$$

$\bar{\varepsilon}_3$ increases drastically with increasing temperature. At T_c a value of 4×10^4 has been observed. Above T_c and down to 100 °C below T_c, $\bar{\varepsilon}_3$ obeys a Curie-Weiss law.

4.3.3 Electrooptic Coefficients

The independent linear electrooptic coefficients in orthorhombic barium sodium niobate are r_{13}, r_{23}, r_{33}, r_{42}, and r_{51}. Their low- and high-frequency values are listed in Table 4.6. Singh et al.[61)] determined r_{13}, r_{23}, and r_{33} interferometrically and r_{42}, r_{51} by measuring the induced birefringence in appropriate directions. In all cases the piezoelectric change of sample length is smaller than the experimental uncertainties. As the elastooptic effect of BNN has not been measured yet, we are not able to calculate r_{ij}^{ε} from r_{ij}^{σ}. The high-frequency constants were determined by Turner[61)] at approximately 100 MHz. In both kinds of experiments the results for r_{13} and r_{23} are nearly the same, indicating again the practically tetragonal symmetry of BNN.

Table 4.6. Electrooptic coefficients of $Ba_2NaNb_5O_{15}$ at $\lambda = 633$ nm; unit 10^{-12} m/V

	r_{13}	r_{23}	r_{33}	r_{42}	r_{51}
r_{ij}^{σ}	15	13	48	92	90
r_{ij}^{ε}	7	8	29	79	95

4.4 Perovskites

4.4.1 Phase Transitions

The general chemical formula of perovskites is ABC_3. In the original perovskite, $CaTiO_3$, C is oxygen. Although structural phase transitions occur in nearly all perovskites, in those compounds with the electrooptically attractive paraelectric-ferroelectric phase transition, C = O. The typical arrangement of atoms in perovskites is outlined in Fig. 4.5a. The central cation B is usually of small size. For the sake of completeness we enumerate some perovskites without oxygen-octahedra: $CsPbCl_3$, $CdMnF_3$, $RbPbBr_3$.

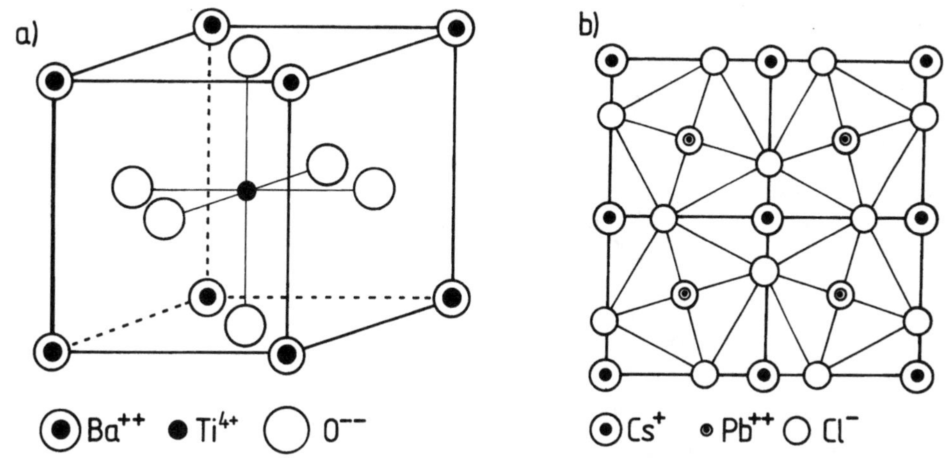

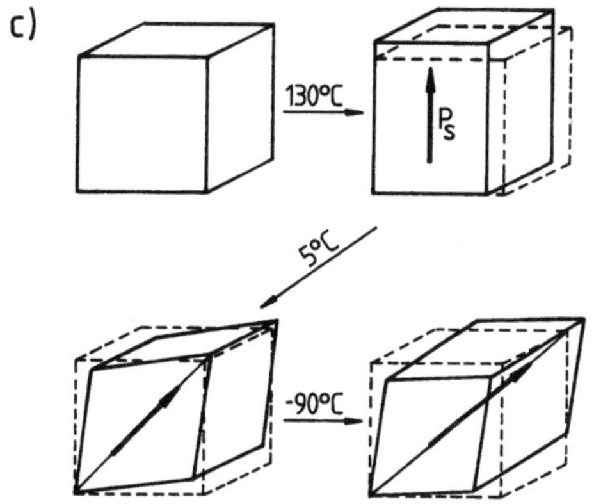

Fig. 4.5 a–c. Different modifications and phase transitions in perovskites. **a** Cubic perovskite-type structure ABC_3. Examples: $BaTiO_3$, $CsPbCl_3$. **b** Structural phase transition from cubic to tetragonal symmetry in $CsPbCl_3$. **c** Ferroelectric modifications of $BaTiO_3$. Point group symmetries are: cubic – tetragonal – orthorhombic – rhombohedral

They show typical structural phase transitions caused by a twisting of the octahedra as sketched in Fig. 4.5b. This kind of twist-like transitions may always be expected in perovskites. They are connected with an enlarging of the unit cell and an orientational change of the crystallographic axes. The typical deformations in oxygen-octahedra crystals which result in ferroelectric unit cells are demonstrated in Fig. 4.5c with $BaTiO_3$. The macroscopic polarization is caused by a movement of the small central cation relative to its octahedral oxygen frame.

Among the perovskites two kinds of compounds are of most importance:

α) $A^{2+}B^{4+}O_3$: $BaTiO_3$; $SrTiO_3$, $PbTiO_3$

β) $A^{+}B^{5+}O_3$: $KNbO_3$, $KTaO_3$.

In all modifications below 130 °C $BaTiO_3$ is ferroelectric. As shown in Fig. 4.5c the spontaneous polarization $\vec{P}_s$ is aligned along [100], [110], or [111]. The directions are assigned to the cubic modification at high temperatures. $SrTiO_3$ suffers a phase transition at 105 K of the twisted type. In $PbTiO_3$ the paraelectric-ferroelectric (cubic-tetragonal) transition occurs at 490 °C[35].

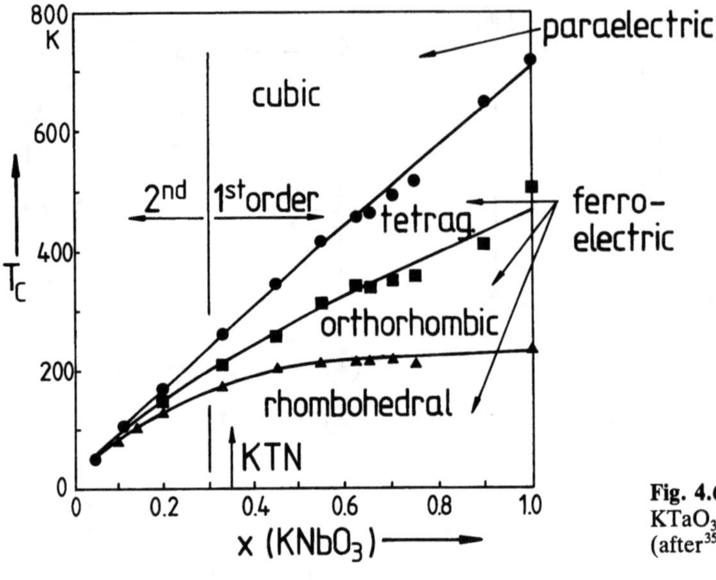

Fig. 4.6. Phase transitions in $KTaO_3$-$KNbO_3$ mixed crystals (after[35])

Cubic $KNbO_3$ and $KTaO_3$ have nearly the same size of unit cells and are miscible in all concentrations. In spite of chemical and structural similarity the paraelectric-ferroelectric transition occurs in $KTaO_3$ at 4 K, but in $KNbO_3$ at 698 K. With $KNbO_3$ the same sequence of phase transitions as with $BaTiO_3$ have been observed[35]. Figure 4.6 exhibits the change of the transition temperature T_c with different compositions of solid solutions. The compound with about 35 mol% $KNbO_3$, called KTN, has a T_c at 283 K. Therefore its dielectric anomalies are very large at room temperature.

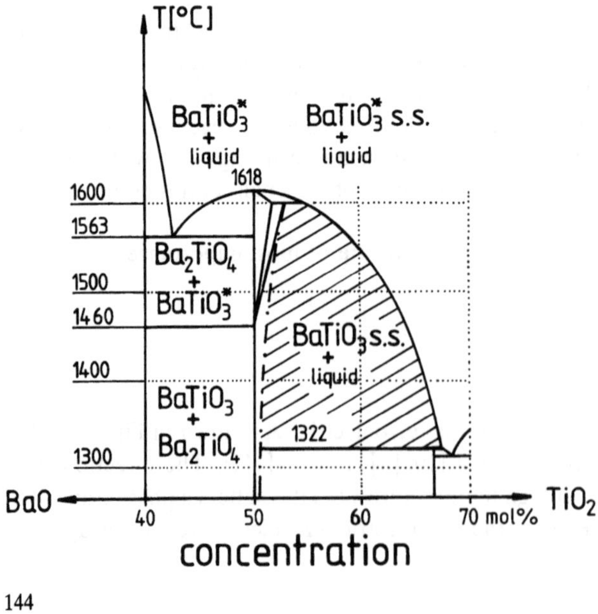

Fig. 4.7. A part of the phase diagram BaO-TiO_2. The hexagonal modification of barium titanate is marked by an asterisk. $BaTiO_3$ without an asterisk presents the cubic modification. Only in the hatched region cubic $BaTiO_3$ can be grown from the melt

4.4.2 *Crystal Growth*

The binary phase diagram TiO_2-BaO (Fig. 4.7) shows the existence of a hexagonal modification of $BaTiO_3$ above 1460 °C. This fact prevents the growth of cubic single crystals from a stoichiometric melt. Different methods have been applied to avoid the hexagonal phase. As an example the addition of 3 mol% $SrTiO_3$ stabilizes the cubic modification. With this composition of the melt Tien and Garretson[62)] grew single crystals of 30 mm length and 5 mm in diameter employing the floating zone technique, which has the advantage that no container is needed. This technique works also with the addition of $CaTiO_3$ instead of the strontium compound. Another method is the growth from a KF-flux, but the endproduct is contaminated by 0.3 wt% F and 0.04 wt% Pt from the crucible[63)]. The phase diagram indicates a possible crystallization of the cubic modification from a TiO_2-rich melt. The temperature of the eutecticum is 1322 °C and to get a solid solution with low Ti concentration, the melting temperature should be as low as possible. From a melt with TiO_2 excess Behruss et al.[64)] pulled with a seed a single crystal of 1 cm^3 volume at 1390 °C. They also produced $SrTiO_3$ crystals, but at a temperature of 1550 °C. Due to a small temperature gradient near the growth interface crystals grew with high perfection. Whereas the concentration of dislocation in a floating-zone $SrTiO_3$ crystal was $10^6 - 10^7$ per cm^2, it was less than 10 per cm^2 in a top-seed grown sample[64)]. Multidomain $BaTiO_3$ can be converted to single domain samples when they pass the paraelectric-ferroelectric phase transition under the action of a dc-field of 10 KV/cm.

Single crystals of $KTaO_3$[65)] and of $KNbO_3$[66)] grow from a K_2O-rich melt by the Kyropoulos- and by the top-seed-technique. Using an oxygen atmosphere $KTaO_3$ becomes colourless and its electrical resistance is high[65)]. Single crystals of 20 × 20 × 5 mm^3 have been obtained[67)]. The quality of $KNbO_3$-crystals depends critically on the amount of K_2CO_3 excess in the melt. Optimal results were achieved with 52.5 mol% K_2CO_3 and with [001]-seeds, where the orientation is referred to the cubic high temperature modification[68)].

Single crystals of 40 × 40 × 15 mm^3 volume exhibited a spatial variation of refractive indices of less than 10^{-5}[69)]. KTN-crystals pulled from the melt usually exhibit striations perpendicular to the growth direction. They are produced by locally varying values of the spontaneous polarization $\vec{P}_s$ as a consequence of an inhomogeneous Ta/Nb-distribution[70)]. The poor perfection of KTN single crystals restricts their application in devices.

4.4.3 *Temperature Dependence of Electrooptic Effects*

The interesting point in these crystals is the critical divergence near the phase transition temperature T_c of all quantities which are related to the electrical polarization $\vec{P}$. In Devonshire's theory[71)] all anomalies of macroscopic properties near a ferroelectric phase-transition are traced back to the occurance of a spontaneous polarization $\vec{P}_s$ at T_c. Therefore the suitable electrooptic coefficients are those for $\vec{P}$, defined as

$$\Delta a_{ij} = g_{ijkl} P_k P_l \,. \tag{4.1}$$

The quadratical dependence has been used because the point group above T_c is centrosymmetric (m3m). The g_{ijkl} in (4.1) are assumed to be practically independent of temperature near T_c in contrast to the R_{ijkl} in

$\Delta a_{ij} = R_{ijkl}E_kE_l$.

The relation between $\vec{E}$ and $\vec{P}$ is given by

$$E = \frac{1}{\varepsilon_0 C}(T - \theta_p)P + \xi P^3 + \zeta P^5 \quad (4.2)$$

for a transition of first order. For a phase transition of second order only two terms are needed in the expansion of $\vec{E}$ as a function of $\vec{P}$. Above T_c, $P = 0$ for $E = 0$, but below T_c the alignment of electrical dipoles establishes a macroscopic polarization $\vec{P}$. From Eq. (4.2) the reciprocal susceptibilities

$$\chi_P^{-1} = \frac{1}{\varepsilon_0 C}(T - \theta_P) \text{ and} \quad (4.3)$$

$$\chi_f^{-1} = \frac{1}{\varepsilon_0 C}(T - \theta_P) + 3\,\xi P_s^2 + 5\,\zeta P_s^4 \quad (4.4)$$

are calculated for the paraelectric and the ferroelectric phase, respectively. P_s is determined as

$$P_s^2 = -\frac{\xi}{2\,\zeta} + \sqrt{\left(\frac{\xi}{2\,\zeta}\right)^2 - \frac{T - \theta_P}{\varepsilon_0 \cdot C\zeta}} . \quad (4.5)$$

According to Eqs. (4.3)–(4.5) the parameters C, ξ, ζ and the Curie-temperature θ_p can be determined with dielectric measurements. For $BaTiO_3$ the numerical values are $C = 1.7 \times 10^5$ K, $\xi = -5.5 \times 10^8\ Vm^5C^{-3}$, $\zeta = 1.7 \times 10^{10}\ Vm^9C^{-5}$, and $\theta_p = 388$ K[35]. Assuming an alignment of $\vec{P}_s$ along the cubic [001]-direction, the optical properties of $BaTiO_3$ are described by

$$a_{11} = a_{11}^0 + g_{1122}(P_s + \chi E_3)^2 \text{ and}$$

$$a_{33} = a_{11}^0 + g_{1111}(P_s + \chi E_3)^2 .$$

Together with Eqs. (4.3) to (4.5) these two relations give the Kerr-effect above T_c and the birefringence and the linear electrooptic effect below T_c. The measurement of only one of these properties at only one temperature is sufficient to obtain Δn, r_{ijk}, and R_{ijkl} for a greater range of temperature. Using $g_{11} - g_{12} = 0.13\ m^4/C^2$[72] and $n = 2.38$ these quantities are presented in Fig. 4.8.

The strong increase of R_{ijkl} just above T_c and the large values for the r_{ijk} below T_c demonstrate the attractiveness of a ferroelectric material for electro-optic applications. Similar considerations are also valid for $LiNbO_3$, tetragonal bronzes and KDP-crystals. Theoretical analyses of linear and nonlinear optical properties of perovskites have been reported mainly by Wemple and Di Domenico[73], including detailed considerations about their application as electrooptic devices[74]. Recently is has been shown that $KNbO_3$ single crystals can be used as waveguide modulators[75].

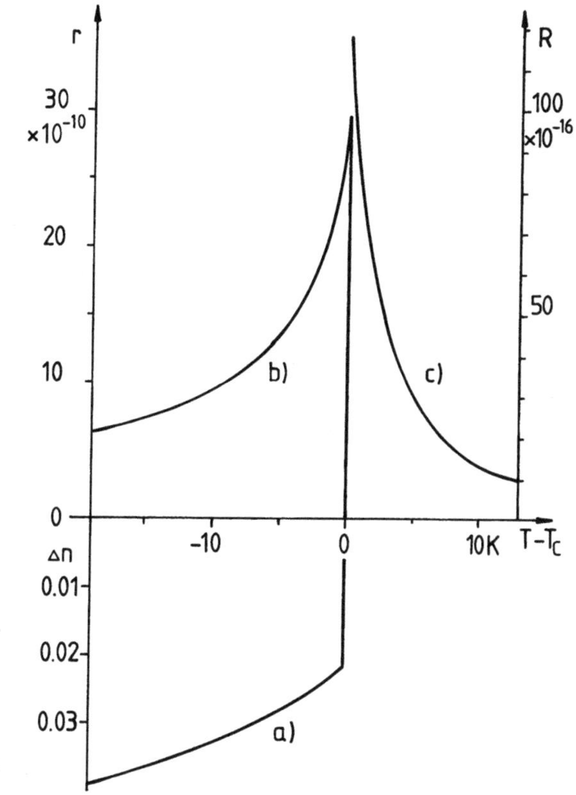

Fig. 4.8. Temperature dependence of electrooptical effects near the paraelectric-ferroelectric phase transition in $BaTiO_3$ at about $T_c = 130\,°C$. **a** birefringence $\Delta n = -0.5\, n^3\, (g_{11} - g_{12})P_s^2$; **b** linear electrooptic coefficient $r = 2(g_{11} - g_{12})P_s\chi_f$; **c** Kerr effect $R_{11} - R_{12} = (g_{11} - g_{12})\chi_p^2$

4.5 KDP-Family

4.5.1 Characterization of Crystals

KH_2PO_4 (KDP) is one of the most extensively investigated electrooptic materials. Large single crystals can be grown at low cost by lowering the temperature or by slow evaporation of a saturated aqueous solution. The preferred growth direction is the crystallographic c-axis and the morphology exhibits a tetragonal prism with pyramidal (101)-faces at the top. The prism faces are stepped faces, which renders the preparation of well orientated samples sometimes difficult. At room temperature the point symmetry group of KDP is $\bar{4}2m$. At low temperatures it changes to the ferroelectric modification 4m. Thus KDP is an example of a non-centrosymmetric crystal above a paraelectric-ferroelectric phase transition. In such a case large piezoelectric and linear electrooptic effects are observed because of their coupling to the dielectric anomaly. The origin of ferroelectricity is the ordering of the protons in a double-well potential. Consequently the transition temperature increases when the proton is replaced by the heavier deuterium. Crystal growth of KD_2PO_4 (DKDP or KD*P) is more complicated than of KDP because there exists a polymorphic monoclinic modification of DKDP. Large single crystals of high optical quality are obtained only in a limited region of temperature and D_2O-concentration. Recently Minhua et al.[76] have reported the optimal conditions for growing DKDP, including the growth in a metastable region. KDP is the most prominent member of a whole family of crystals. H can be replaced by D, PO_4 by AsO_4, and K by NH_4, Rb, and Cs. Although the electrooptic coefficients of ADP and CDA, for instance are larger than those of KDP, we will describe the latter compound more extensively because most physical data are available about this crystal. Several members of the KDP-family are widely used as Pockels-cells and frequency-doublers in laser technology.

4.5.2 Measurements of Mechanically Free Coefficients

The matrices of piezoelectric, linear electrooptic, and elastooptic constants for point group $\bar{4}2m$ are presented in Table 4.7. The natural choice of the experimental method for determining r_{63}^{σ} is the measurement of field induced birefringence $\Delta\delta$ in the transversal configuration $\vec{k} \parallel [110]$, $\vec{E} \parallel [001]$ or in the longitudinal configuration $\vec{k} \parallel \vec{E} \parallel [001]$ (Fig. 4.9). The induced birefringences are expressed by

$$\Delta\delta = \frac{\pi}{\lambda_0} [n_0^3 r_{63} - 2(n_e - n_0) d_{36}] L \cdot E \tag{4.6a}$$

for the transversal and by

$$\Delta\delta = \frac{2\pi}{\lambda_0} n_0^3 r_{63} L \cdot E \tag{4.6b}$$

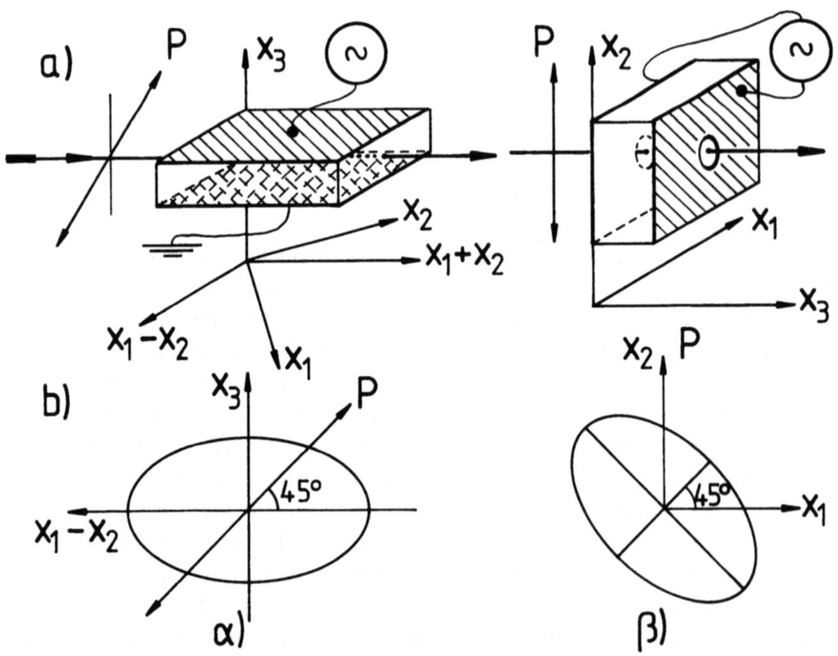

Fig. 4.9 a, b. Transversal (α) and longitudinal (β) configuration for measuring the electrooptic coefficient r_{63} in KDP. **a** Relative orientation of sample, electric field, and light beam. **b** Orientation of the polarizer P and the natural (α) or induced (β) birefringence

Table 4.7. Matrices of electrooptic (r_{ij}), piezoelectric (d_{ij}) and elastooptic (p_{ij}) coefficients in point group $\bar{4}2$ m

0	0	0
0	0	0
0	0	0
r_{41}	0	0
0	r_{41}	0
0	0	r_{63}

0	0	0	d_{14}	0	0
0	0	0	0	d_{14}	0
0	0	0	0	0	d_{36}

$\mathcal{P}_{11}$	$\mathcal{P}_{12}$	$\mathcal{P}_{13}$	0	0	0
$\mathcal{P}_{12}$	$\mathcal{P}_{11}$	$\mathcal{P}_{13}$	0	0	0
$\mathcal{P}_{31}$	$\mathcal{P}_{31}$	$\mathcal{P}_{33}$	0	0	0
0	0	0	$\mathcal{P}_{44}$	0	0
0	0	0	0	$\mathcal{P}_{44}$	0
0	0	0	0	0	$\mathcal{P}_{66}$

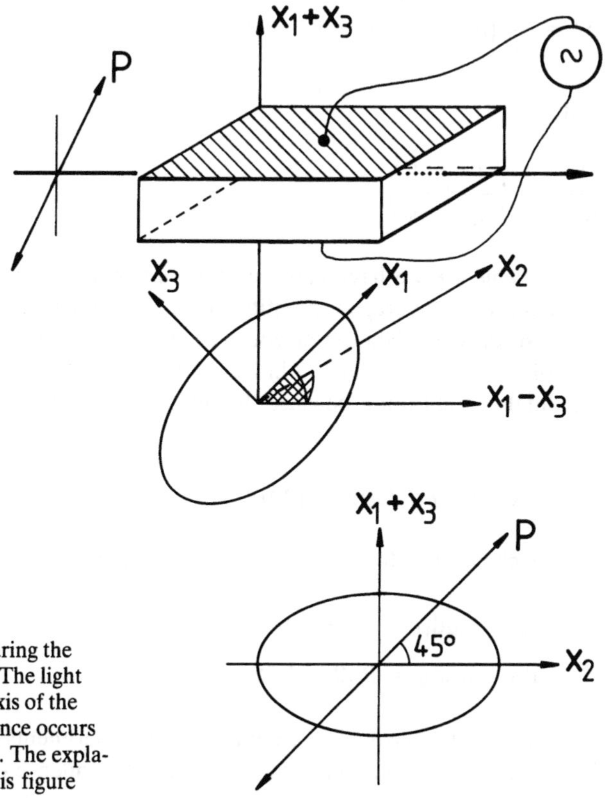

Fig. 4.10. Configuration for measuring the electrooptic coefficient r_{41} in KDP. The light beam is not parallel to a principal axis of the indicatrix. Therefore ray-birefringence occurs as already demonstrated in Fig. 2.1. The explanations of Fig. 4.9 apply also to this figure

Table 4.8. Numerical values of electrooptic (r_{ij}), piezoelectric (d_{ij}), and elastooptic coefficients of KDP for room temperature and a wavelength of 633 nm

$r^{\sigma}_{41} = 8.7 \times 10^{-12}$ m/V	$d_{14} = 1.28 \times 10^{-12}$ C/N
$r^{\sigma}_{63} = 10.3 \times 10^{-12}$ m/V	$d_{36} = 20.90 \times 10^{-12}$ C/N
$p_{11} = 0.266$	$p_{31} = 0.230$
$p_{12} = 0.261$	$p_{33} = 0.17$
$p_{13} = 0.207$	$p_{44} = -0.02$
	$p_{66} = -0.054$

for the longitudinal arrangement. By the same method r^{σ}_{41} is determined with a $[101]'$,$[10\bar{1}]'$,$[010]'$-specimen as illustrated in Fig. 4.10. The primed directions refer to cartesian axes x'_i. Their directions, but not their lengths, coincide with the crystallographic ones. For $\Delta\delta$ we obtain

$$\Delta\delta = \frac{\pi}{\lambda_0}\,[n'^3_e r_{41} - 2\,(n'_e - n_0) d_{14}] L \cdot E \tag{4.6c}$$

with

$$\frac{1}{n'^2_e} = \frac{1}{2}\left(\frac{1}{n^2_e} + \frac{1}{n^2_0}\right).$$

In Table 4.8 we have gathered the numerical values for the r_{ij}, d_{ij}, and p_{ij} of KDP. They represent averaged quantities of a lot of coefficients reported in literature[28)]. On the basis of these data the piezoelectric change of sample length amounts to only 3% in Eq. (4.6a) and less than 1% in Eq. (4.6c), whereas it is missing completely in the longitudinal configuration because the light wave propagates along the optical axis.

4.5.3 Clamped Coefficients

The mechanically clamped electrooptic constants are calculated to be (unit: 10^{-12} m/V)

$$r^{\varepsilon}_{41} = r^{\sigma}_{41} - p_{44} d_{14} = 8.6 + 0.03 = 8.6$$

$$r^{\varepsilon}_{63} = r^{\sigma}_{63} - p_{66} d_{36} = 10.0 - 1.25 = 8.75\ .$$

These values have to be compared with experiments performed at high frequencies. Pisarevskii et al.[77)] measured the frequency dependence of r_{63} between 10 and 10^{10} cycles per second with a cylindrically shaped sample in a coaxial cavity resonator. The optical axis of the crystal coincided with the geometrical axis of the cylinder. They observed a piezoelectric resonance at about 9×10^4 Hz. Below and above this resonance the electrooptic response is essentially free from any dispersion. At high frequencies the magnitude of r_{63} is reduced by about 18% to 8.5×10^{-12} m/V. Pursey et al.[78)] measured

$$r_{63} = 10.2 \times 10^{-12}\ \text{m/V at } 9.17 \times 10^9\ \text{Hz}$$

for an optical wavelength of 633 nm. Employing the heterodyne technique at 10^7 c/s on three different samples, Rosner et al.[79] observed the average value $r_{63} = 8.8 \pm 0.5 \times 10^{-12}$ m/V. With the pulse-technique Veeser et al.[26] determined $r^{\sigma}_{63} = 10.0 \pm 0.6 \times 10^{-12}$ m/V and $r^{\varepsilon}_{63} = 8.5 \pm 0.4 \times 10^{-12}$ m/V. This feature of $r_{63}(r^{\varepsilon}_{63} < r^{\sigma}_{63})$ is observed with all compounds of the KDP-family. Whereas the decrease of about 1.4×10^{-12} m/V is the same in DKDP and in RDP, it is a little bit smaller in KDA and RDA, and a little bit larger in ADP and AD*P[80].

4.5.4 Influence of Deuteration

Volkova and Velichko[81] studied extensively the influence of deuteration on the electrooptical properties of KDP. Their results are presented in Fig. 4.11. Whereas r_{63} increases from KDP to DKDP, $r_{63}/\varepsilon_0(\tilde{\varepsilon}_r - 1)$ is nearly a constant. Thus the dependence of r_{63} on deuteration is caused by the increase of the dielectric constant $\tilde{\varepsilon}_r$, which again is a consequence of the increasing transition temperature.

4.5.5 Dispersion

In KDP the half-wave voltage V_π increases nearly linearly with λ. In contrast to this behaviour the electrooptical coefficient r_{63} at first increases with increasing photon

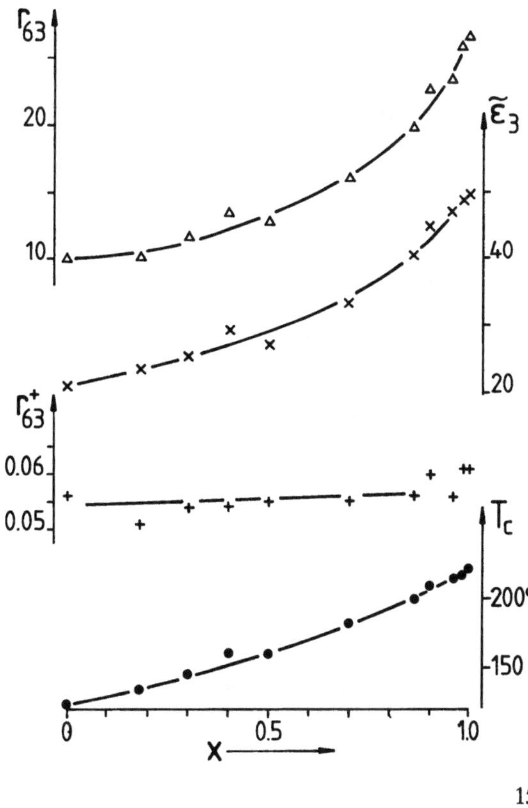

Fig. 4.11. Influence of deuteration on the electrooptic coefficient r_{63} of $K(D_xH_{1-x})_2PO_4$. $r^{+}_{63} = r_{63}/\varepsilon_0(\tilde{\varepsilon}_3 - 1)$. Units of r_{63} and r^{+}_{63} are 10^{-12} m/V and m^2/C, respectively. After[81]

energy but above 6 eV it gets smaller[82]. Thus the dispersion of V_π is strongly influenced by the dispersion of n. The difference in optical and electrooptical dispersion may be vizualised as follows. In a similar way the Kramers-Kronig dispersion relation

$$\varepsilon_{ij,1} - \delta_{ij} = \frac{2}{\pi} \mathcal{P} \int \frac{\omega' \varepsilon_{ij,2}}{\omega'^2 - \omega^2} \, d\omega'$$

connects the real part $\varepsilon_{ij,1}$ of the dielectric constant to its imaginary part $\varepsilon_{ij,2}$,

$$\Delta\varepsilon_{ij,1} = \frac{2}{\pi} \mathcal{P} \int \frac{\omega' \Delta\varepsilon_{ij,2}}{\omega'^2 - \omega^2} \, d\omega' \tag{4.7}$$

does the same for the increments. In Eq. (4.7) $\mathcal{P}$ denotes the principal value of the integral. Only the changes of oscillator strengths, but not of resonance frequencies have been taken into account. By the dispersion relation (4.7) the origin of the electrooptic response in the transparent region is traced back to the $\Delta\varepsilon_{ij,2}$-spectrum. Onaka and Ito[83] observed in the VUV- and XUV-reflection-spectrum several absorption bands, the first one lying at about 8 eV. Comparing the spectra of KDP and ADP, they related the bands up to 20 eV to transitions within the ions PO_4, H_2PO_4, or H_4PO_4 and assigned the first two resonances at 8 eV and 9 eV to transitions from O_π-states to 1s hydrogen and 2p oxygen orbitals, respectively. If the relative magnitudes $\varepsilon_2(8\text{ eV})/\varepsilon_2(9\text{ eV})$ and $\Delta\varepsilon_2(8\text{ eV})/\Delta\varepsilon_2(9\text{ eV})$ differ, then also the optical and electrooptical dispersion must be different. Furthermore the resonances relevant for the electrooptic response are centered at lower energies than those which are responsible for the refractive indices.

4.6 Semiconductors

4.6.1 Crystallography

The most common semiconductors are Si and Ge, crystallizing in the diamond structure shown in Fig. 4.12. It consists of two face centered lattices, shifted [1/4, 1/4, 1/4] towards each other. In the sphalerite-structure one sublattice is occupied by a cation of the third or second column and the second one by an anion of the fifth or sixth column in the table of chemical elements. The typical structural feature of these compounds is their tetrahe-

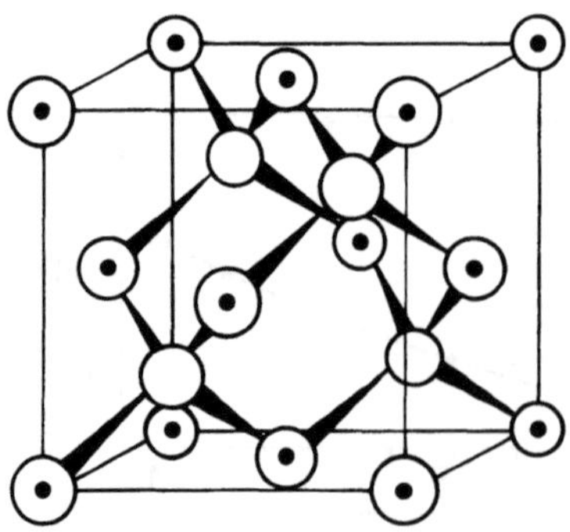

Fig. 4.12. Cubic ZnS (sphalerite) structure. In the diamond structure both kinds of atoms are the same

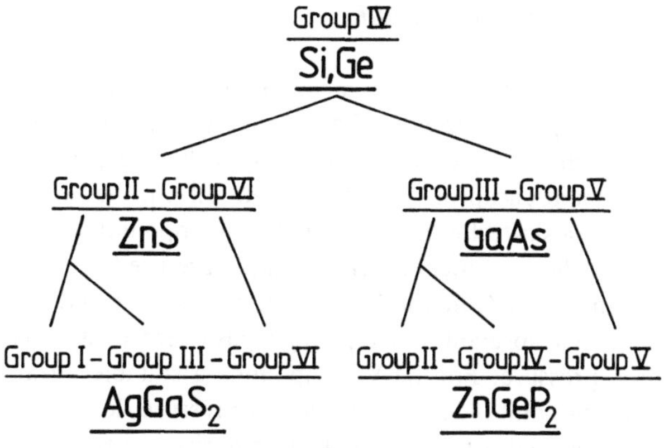

Fig. 4.13. Schematic illustration of the formation of chalcopyrites

dral coordination. The different constituents reduce the symmetry in III-V-, II-VI- and I-VII-compounds from m3m to $\bar{4}3m$ and thus they possess a linear electrooptical effect. In a regular tetrahedron, however, the structural acentricity is only small and consequently the electrooptic coefficients are not too large. This deficiency is compensated by large refractive indices, resulting in acceptable modulation properties especially in the infrared region.

Figure 4.13 illustrates the potentiality to produce new materials by increasing the chemical variety. They are called chalcopyrites. As shown in Fig. 4.14 their tetragonal structure is derived from tow unit cells of the cubic zincblende. In most compounds the c-axis is slightly compressed and c/a is smaller than two [84]. The point symmetry is $\bar{4}2m$ and in several compounds the birefringence enables phase matching in nonlinear optical applications. Some AB-compounds like CdS, CdSe, and ZnS crystallize in the hexagonal wurtzite structure having point symmetry 6mm[85].

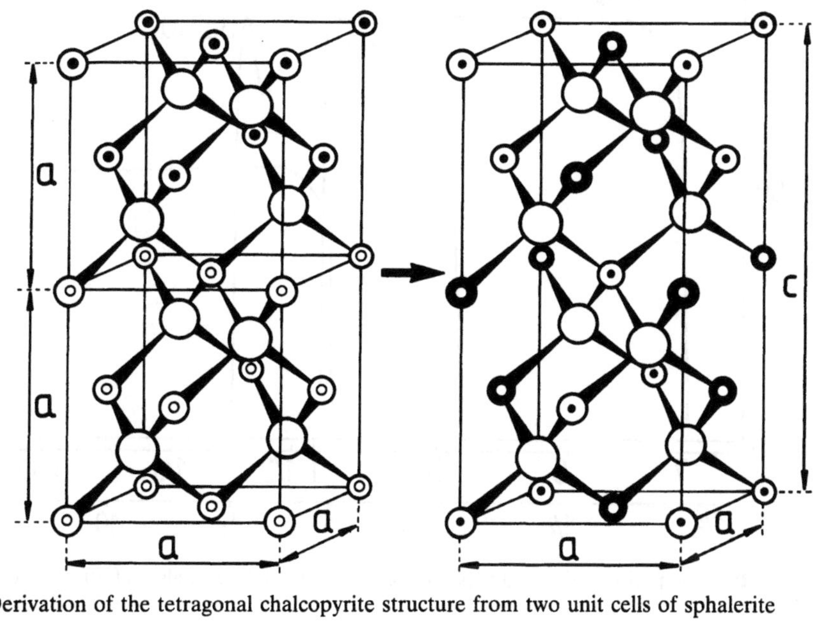

Fig. 4.14. Derivation of the tetragonal chalcopyrite structure from two unit cells of sphalerite

4.6.2 Crystal Growth

In contrast to Si and Ge most compound semiconductors contain volatile constituents, which rapidly evaporate. Therefore crystal growth has to be performed in closed systems, either in a completely sealed set-up or by encapsulating the melt. For III-V-compounds the liquid encapsulated Czochralski-(LEC), the horizontal Bridgman- and the magnetic Czochralski-technique and for II-IV-compounds the horizontal Bridgman- and the vertical Bridgman-Stockbarger-technique are used[86].

In the LEC-method votalization is suppressed by covering the melt surface with a liquid encapsulant and applying a gas overpressure that exceeds the compound dissociation pressure. For III-V-compounds a suitable encapsulant is dry B_2O_3. It has a low melting point (450 °C), low vapour pressure, low density, good optical transparency and it is immiscible with the melt. In a horizontal Bridgman apparatus crystal growth can be started with a seed, and adding an excess of the volatile constituent (e.g. As in growing GaAs) ensures stoichiometry of the melt. With both techniques GaAs, GaP, and InP have been grown[86]. Several II-VI-compounds, like CdSe and CdTe, are obtained by the vertical Bridgman-Stockbarger-method[87]. ZnS, ZnSe, CdS, and ZnTe require additionally high pressure autoclaves[88]. In the magnetic Czochralski-technique growth takes place in a sealed puller ampoule. The seed holder has a core of magnetic material and can be lifted by a magnet outside the tube. In this way InAs and GaAs have been prepared[89].

In all methods mentioned the ampoules are made of quartz and the material of crucibles is also quartz or graphite for allowing RF-heating.

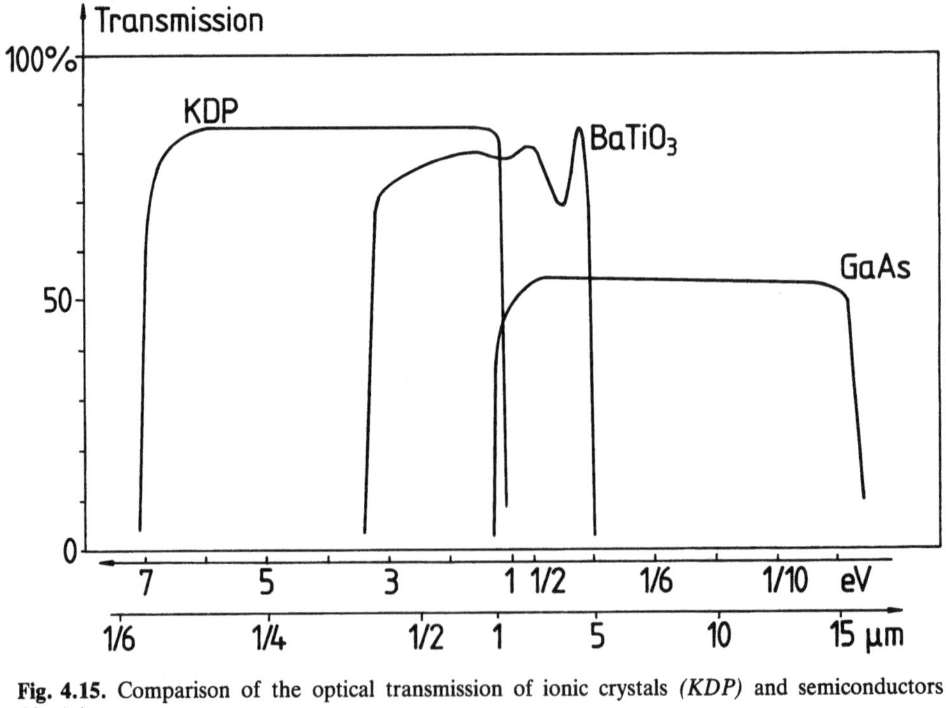

Fig. 4.15. Comparison of the optical transmission of ionic crystals *(KDP)* and semiconductors *(GaAs)*

Problems in the growth of chalcopyrites are in general similar to those of AB-compounds. Additional difficulties arise from pecularities in the phase-diagrams. Crystallographic data, crystal growth conditions, and physical properties of chalcopyrites have been summarized by Shay and Wernick[84)].

Semiconductors are also in use as waveguides. One method of their fabrication is the preparation of heterostructures by sandwiching a GaAs layer between two layers of AlGaAs. The refractive index of $Al_xGa_{1-x}As$ is less than that of GaAs by about $\Delta n = -0.4\,x$ and total internal refraction occurs at both interfaces[39)].

If one-dimensional periodic structures are produced with ultrathin alternating GaAs-$Al_xGa_{1-x}As$-layers they are called multiple quantum wells or superlattices. In the last fifteen years epitaxial growth techniques have been developed with such a success that layers as thin as 10 nm could be prepared. The most important technique is the molecular-beam-epitaxy (MBE), which is, in principle, a very simple process. Under ultra-high-vacuum conditions ($< 10^{-10}$ torr) a beam of molecules is directed at a substrate surface. The sticking coefficient, which measures the ratio of desorbed to incident flux intensity, is mainly influenced by the temperature of the substrate and by the relative flux intensities of the crystal constituents. The effusion cells, where the molecular beams are produced by heating, can be closed by shutters. Thus it is possible to control the composition of the surface at each growth step. Usually the growth chamber is provided with instruments for structural and compositional analysis of the surface. Review articles of several groups which are active on the field of MBE-growth, are gathered in a book edited by Chang and Ploog[90)].

4.6.3 *Optical Properties*

In Fig. 4.15 the transmission regions of semiconductors are compared with those of other types of electrooptic crystals. Obviously semiconductors are favourable materials in the infrared region, because their lattice vibrations, causing the long-wavelength absorption bands, are excited at much lower energies than the intermolecular vibrations in molecular units, such as PO_4 in KDP or TiO_6 in perovskites. The importance of infrared optics stems from laser technology, communication technology and military applications. At the high energy side the transparency of a material is limited by the band gap. In experiments with semiconductors the energy of the probe beam is usually near to the band energy. In this region all optical constants are enhanced due to the vicinity of the optical resonance. Another factor which influences strongly the magnitude of optical constants is the nature of the band gap. Generally spoken, oszillator strengths are larger in direct allowed transitions than in dipole forbidden or in indirect ones. Near the band gap the shape of absorption spectra depends on the occurence of excitons and on the concentration of impurities. Therefore it may differ from sample to sample. In the long wavelength tail photoconduction effects can reduce the strength of external electrical fields and thereby also the electrooptical response.

There exists a marked difference between the absorption spectrum of GaAs bulk crystals and of a GaAs-AlAs superlattice. Whereas in the bulk material narrow exciton peaks are observed only at low temperatures, the superlattice spectrum shows the excitonic structure even at room temperature[91)]. The refractive index of a superlattice is not solely determined by the average of the Al content, but depends strongly on the barrier layer thickness[92]. Furthermore superlattices are birefringent: for light polarization parallel to the layers the index of refraction is larger than for the perpendicular direction[93].

4.6.4 Electrooptical Effect

In point group $\bar{4}3m$ only one independent electrooptical coefficient exists: $r_{41} = r_{52} = r_{63}$. It is possible to determine r_{41} with several different configurations. Usually the light beam is chosen transverse to the electric field. Assuming that the beam is aligned along $[1\bar{1}0]$, the induced birefringence

$$\Delta\delta = -\frac{\pi}{\lambda} n^3 r_{eff} \frac{L}{d} V$$

depends strongly on the orientation of the electric field. This is designated by the angle χ in Fig. 4.16a. Distinguishing between the two cases i) birefringence parallel to the elec-

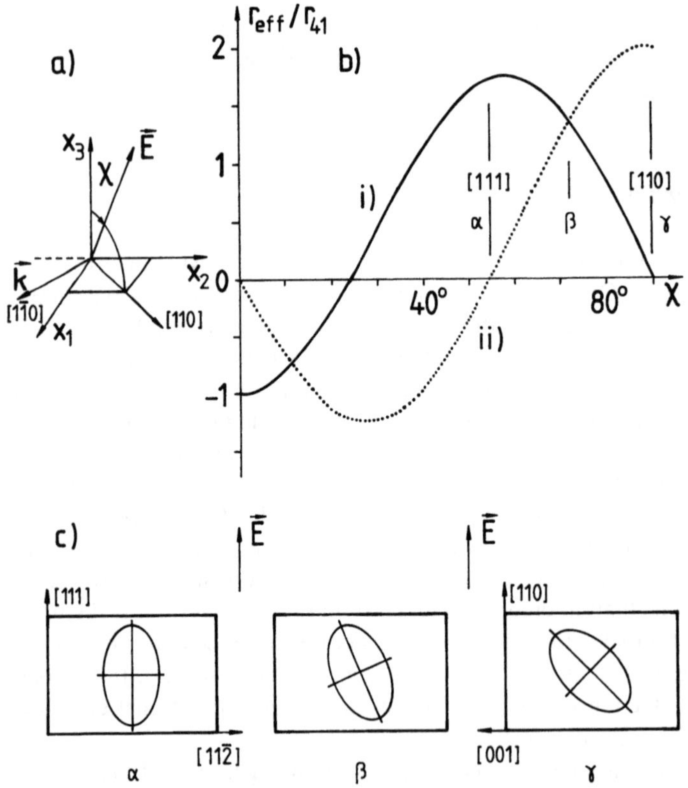

Fig. 4.16 a–c. Transversal electrooptical effect in crystals of point group $\bar{4}3m$ as a function of the electric field direction. **a** Relative orientation of $\vec{k} \parallel [1\bar{1}0]$ and $\vec{E}$. **b** The total induced birefringence $\Delta\delta$ is decomposed into a part (i) parallel to $\vec{E}$ and a part (ii) inclined 45° to $\vec{E}$. **c** Orientation of the induced birefringence for the three different cases indicated in **b**

tric field and ii) birefringence rotated by 45° relative to the electric field $\vec{E}$, we have calculated r_{eff} as shown in Fig. 4.16b. Notice, that for an oblique direction of $\vec{E}$ the polarization directions of the light wave have no simple orientation relative to $\vec{E}$. This fact renders measurements difficult and it is convenient to realize the configuration $\chi = 0$ or $\chi = 90°$. In the latter case the maximal effect is obtained and the electric field intersects both polarization directions of the light wave (Fig. 4.16c). The relation between mechanically free and clamped coefficients are given by the simple relation

$$r_{41}^{\sigma} - r_{41}^{\varepsilon} = p_{44}d_{14} .$$

In measuring the electrooptic response of a semiconductor several false effects may lead to uncorrect results. Firstly photoconduction can reduce the effective electric field as already discussed.

Secondly dc-conductivity may lead to space charges and as a consequence the effect depends on the position of the light beam and on the kind of the electrode. Thirdly, although the crystals are cubic, strains usually induce a natural birefringence, which interferes with the electrically induced ones. Therefore we add some experimental details in the following survey.

Among the III-V-compounds the electrooptical effect has been measured in GaP, GaAs and InP[85]. Nelson and Turner[94] used GaP crystals of high resistivity (10^7 Ω-cm) and observed by the pulse method the constant-strain values r_{41} between 0.56 and 3.39 μm. Within the experimental accuracy $r_{41}^{\varepsilon} = -1.02 \pm 0.05 \cdot 10^{-12}$ m/V does not vary with the laser wavelength. They also found that $r_{41}^{\sigma} < r_{41}^{\varepsilon}$. Sugie and Tada[95] measured r_{41}^{σ} in a chromium doped GaAs sample of about 10^8 Ωcm resistivity at 1.15 μm, 3.34 μm, and 10.6 μm. The results were 1.43, 1.24, and 1.51 $\cdot 10^{-12}$ m/V, nearly equal to the average values of earlier experiments. At 1.15 μm the effect was strongly intensity dependent. Nearly the same coefficient as in GaAs has been observed in InP[96].

Electrooptically interesting cubic II-VI-materials are CdTe, ZnS, and ZnSe. CdTe seems to be a good modulator crystal in the infrared. At 10.6 μm $r_{41} = 6.8 \times 10^{-12}$ m/V[97] is much larger than in GaAs and also the figure of merit (n^3r) is twice that of GaAs. ZnS can be used in the infrared as well as in the visible region. Baldassarre et al.[98] studied melt grown crystals possessing a resistivity of 10^9 Ω cm. For a vakuum wavelength of 0.633 μm they obtained $r_{41}^{\sigma} = -1.9 \times 10^{-12}$ m/V. Practically the same value has been reported for mineral crystals[99] and in crystals grown by the hydrothermal method[100].

Mohler and Thomas[101] analyzed the constant-stress electrooptical effect of the I-VII-compound CuCl in respect to different mechanisms contributing to the total effect. The value of about -5×10^{-12} m/V does not change between 700 nm and 450 nm, but increases drastically just above the absorption edge near 400 nm. Nearly the same dispersion is observed at 80 K but the effect is appreciably reduced. The positive temperature coefficient seems to be characteristic for all crystals with sphalerite structure.

The II-VI-compound CdS crystallizes in the wurtzite-structure. The point symmetry is 6mm and the three nonvanishing independent electrooptic coefficients are $r_{13} = r_{23}$, r_{33}, and $r_{42} = r_{51}$. For $\lambda = 1.15$ μm Sugie and Tada[95] determined the numerical values $r_{13} = 3.1$, $r_{33} = 3.2$, and $r_{42} = 2.0 \cdot 10^{-12}$ m/V. The figure of merit n^3r is increased slightly when 25% of S is replaced by Se[102].

Chalcopyrites have the point symmetry $\bar{4}2m$ and thus the linear electrooptic coefficients are the same as in KDP, namely r_{63} and r_{41}. The effect has been determined in

$AgGaS_2$[103], $AgGaSe_2$[104] and in $ZnGeP_2$ and $CuGaS_2$[105]. In Table 4.9 we have gathered the numerical values. Some of these compounds have an only small birefringence and optical inhomogeneities due to strains and even optical activity may influence the measurements[104].

In a GaAs-AlGaAs-superlattice a linear electrooptic coefficient of $r_{41} = -1.6 \cdot 10^{-12}$ m/V has been measured[106]. This is practically the same as in GaAs bulk crystals. The difference between both kinds of materials is more significant in electroabsorption experiments. Due to a mechanism which is called "quantum-confined stark effect"[107] electroabsorption in superlattices can be greatly enhanced over that in bulk semiconductors[108]. Another difference is the strong polarization-sensitive effect in quantum wells for light waves traveling parallel to the layer[109].

Table 4.9 Electrooptic coefficients of chalcopyrites

		λ	r_{41}	r_{63}
$AgGaS_2$	r^{σ}	0.6328	4.0	3.0
$AgGaSe_2$	r^{σ}	1.15	4.5	3.9
$ZnGeP_2$	r^{ε}	3.39	1.6	–0.8
$CuGaS_2$	r^{ε}	0.6328	1.76	1.35
		1.15	1.9	1.66
		3.39	1.1	1.05

units: λ: μm, r_{ij}: 10^{-12} m/V
r^{σ} mechanically free constants
r^{ε} mechanically clamped constants

5 Devices

5.1 General Considerations

5.1.1 Classification Schemes

The most general type of electrooptic device is a modulator. Independent of the actual application as a switch, deflector or something else, the primary step is always a modulation of the optical phase velocity by an electric field. Some underlying physical ideas for the application of electrooptical effects have already been worked out previously and excellent review articles are available[110–112].

Different schemes for classifying devices are in use and sometimes it is only a matter of taste how a special device is denominated. Due to the relative alignment of the electric field and the optical wavevector longitudinal and transversal modulators can be distinguished (Sect. 5.2). In both configurations phase- or intensity-modulation may be realized, where in the latter case the induced change of phase is converted into an intensity variation in a two-beam interferometer or through polarizers. Although the mechanisms of light modulation in bulk crystals and in integrated optics are the same in principle, some different aspects become relevant in waveguides, which are considered in

Sect. 5.3. Most of the more recently fabricated modulators are waveguides with a time response as short as possible and progress depends mainly on technical details of preparation. Some examples of these developments are included in Sect. 5.3. In Sect. 5.4 bulk- and waveguide-deflectors are considered.

In the last chapter we have gathered different kinds of devices, such as electroabsorptive and bistable modulators and the important spatial light modulators which are based on the photorefractive effect.

In detail the design of a device depends on its application. With this in mind a classification would distinguish between light modulators, switches, filters, beam deflectors, frequency translators, waveguide couplers, analog to digital converters, Fourier transformers, spatial light modulators, and so on.

5.1.2 Efficiency

In intensity modulators I varies with $\cos^2(\phi/2)$ (Chap. 3.2) or with $\sin^2(\delta/2)$ (Chap. 3.3). Thus maximal modulation depth is obtained for

$$\phi(E) = \frac{\pi}{\lambda} n^3 r L \frac{V}{d} = \pi , \qquad (5.1)$$

if $\phi(0) = 0$. V is the applied voltage and d the electrode spacing. The value of V, which fulfills the condition (5.1) is the half-wave voltage

$$V_\pi = \frac{\lambda}{n^3 r} \left(\frac{d}{L}\right) . \qquad (5.2)$$

In longitudinal modulators the geometrical factor $d/L = 1$, but it may become much smaller in the transversal configuration.

Another quantity which characterizes the efficiency of a device is the required reactive drive power P_r. Consider an amplitude modulator that turns the light on and off, which is achieved by the voltage $V = \frac{V_\pi}{2}(1 + \sin \omega t)$ if $\phi(0) = \pi/2$. Then

$$P_r = \omega C V_\pi^2/8 ,$$

when any circuit capacitance and conductance being parallel with the crystal, are neglected. For a rod of square cross section d^2, the capacitance C is

$$C = \varepsilon_0 \varepsilon L$$

and the power per unit bandwidth $\Delta\nu$ reads

$$P_r/\Delta\nu = \frac{1}{4} \pi \varepsilon_0 \varepsilon \frac{\lambda^2}{n^6 r^2} \frac{d^2}{L} \qquad (5.3)$$

for a transverse propagation of light.

V_π (5.2) and $P_r/\Delta v$ (5.3) are reduced by decreasing d and increasing L. This way to enhance the efficiency, however, is limited by diffraction effects. Let us consider a Gaussian laser beam propagating in the lowest order transverse mode. By means of a lense it can be focused to that it just passes through a cylinder of length L. Optimizing the focus in respect to the diameter of the cylinder, the beam diameter is 2 w_0 at the waist and 2 $\sqrt{2}\, w_0$ at the entrance- and exit-face as illustrated in Fig. 5.1. The relation between w_0 and L is given by[110].

$$w_0^2 = \lambda L/2\, \pi n \ .$$

The beam diameter is measured to points where the intensity is reduced to $1/e^2$ of its value on the beam axis. Then the minimum value of d is given by

$$d = S \cdot 2\sqrt{2}\, w_0 = S \cdot 2\sqrt{2}\sqrt{\lambda L/2\, \pi n} \ . \qquad (5.4)$$

For practical reasons the safety factor $S (S > 1)$ is introduced in the above equation. Usually $S = 3$ is assumed to be an experimentally manageable condition[110, 111]. Inserting (5.4) into (5.3) yields

$$P_r/\Delta v = \varepsilon_0 \varepsilon S^2 \frac{\lambda^3}{n^7 r^2} \ .$$

V_π as well as $P_r/\Delta v$ may be regarded as an appropriate quantity for characterizing an electrooptic material in respect to its application.

5.1.3 Bandwidth

If the modulation signal is at very high frequencies the efficiency of a device is reduced and it is necessary to look at the limiting factors. Consider first the situation outlined in Fig. 5.2a. R_s is the source impedance of the generator, C the capacitance of the electrooptic crystal, and $V = V_0 \sin \omega t$ the driving voltage. The magnitude of the impedance of the circuit is

$$|Z| = \sqrt{R^2 + \frac{1}{\omega^2 C^2}}$$

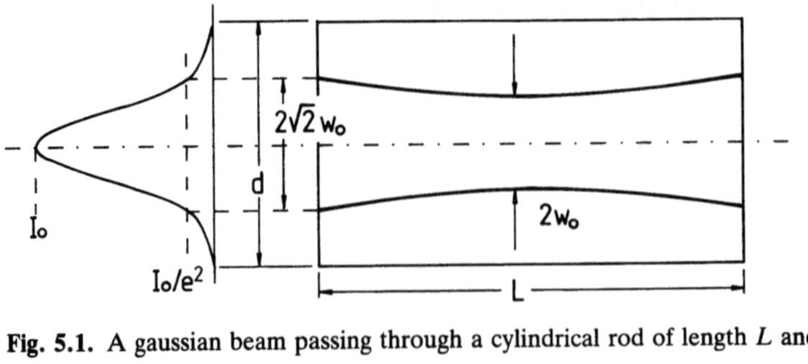

Fig. 5.1. A gaussian beam passing through a cylindrical rod of length *L* and diameter *d*

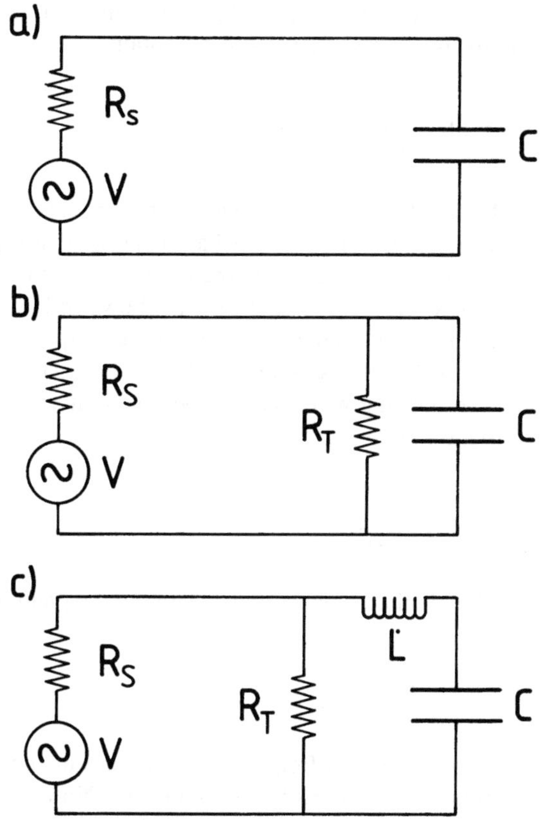

Fig. 5.2 a–c. Circuit models of electrooptic modulators. *C* capacity of the crystal, R_S source impedance of the generator, R_T impedance in which the electrodes are terminated, *L˙* inductance of the electrodes and the bonding wires, *V* driving voltage

and of the capacitor

$$|Z_c| = \frac{1}{\omega C} .$$

Thus the potential drop across the crystal is given by

$$V_c = \frac{V_0}{\sqrt{1 + R^2\omega^2C^2}} .$$

The frequency at which $V_c = V_0/\sqrt{2}$ is called the bandwidth $\Delta\nu$ of the device. In our example

$$\Delta\nu_{RC} = \frac{1}{2\,\pi RC}$$

with $C = \varepsilon_0\varepsilon \frac{WL}{d}$, where W designates the width of the electrodes. Assuming $W = 2\,d$,

$$\Delta\nu = \frac{1}{4\,\pi \cdot R \cdot L} .$$

For a given length L the diffraction limit Eq. (5.4) determines the electrode spacing d and for the half-wave voltage we obtain

$$V_\pi = 2\,S\,\frac{\lambda^{3/2}}{\pi \cdot r \cdot n^{7/2}}\,\frac{1}{\sqrt{L}}\,.$$

Notice that the decrease with increasing L is stronger for $\Delta\nu$ than for V_π. Thus in bulk modulators the price which has to be paid for a small modulation voltage is a low bandwidth. $\Delta\nu$ is doubled when the device is matched to the source by an additional resistor $R_{term} = R_s$ parallel to C as shown in Fig. 5.2b.

In the considered example the charging speed of the capacitor is limited by R_s and C, which is called the RC-limit. At high frequencies also the inductance $L^{\bullet}$ of the electrodes and the bonding wires have to be taken into account (Fig. 5.2c), establishing a second restriction of $\Delta\nu$ by[113)]

$$\Delta\nu_{L^{\bullet}C} = \frac{1.6}{2\,\pi(L^{\bullet}C)^{1/2}}\,.$$

This $L^{\bullet}C$-limited bandwidth is related to the phase velocity v of the modulation signal[113)]:

$$\Delta\nu_{L^{\bullet}C} = 1.6\,\frac{v}{2\,\pi \cdot L}\,.$$

In practice high frequency devices are manufactured as waveguides, where the diffraction limit (Eq. 5.4) is meaningless. Now another effect becomes important. The wavelength λ_m of the modulation field may become comparable with the sample length. Only for

$$\lambda_m \gg 2\,L$$

the electric field strength is homogeneous within the crystal. Devices which fulfill this condition are called lumped modulators. Recently Becker[114)] has measured the frequency dependence of different Mach-Zehnder interferometers, fabricated in $LiNbO_3$. He found that they can be analysed on the basis of a simple $RL^{\bullet}C$ circuit, such as it is illustrated in Fig. 5.2c. Depending on the values of the inductance and of the capacitance he achieved 3 dB bandwidths (this corresponds to a decrease of the signal by a factor of $\sqrt{2}$) from 280 MHz to 2.75 GHz and V_π's from 1 V to 4 V, respectively.

At even higher frequencies the modulators are traveling wave elements which are considered in Sect. 5.3.2.

5.2 Bulk Modulators

5.2.1 Longitudinal Modulators

If the electric field and the optical wavevector are aligned in the same direction, the induced phase difference is independent of geometrical dimensions:

$$\Delta\phi = \frac{\pi}{\lambda} n^3 r V .$$

V is the applied voltage and r the actual electrooptic coefficient. The advantage of this configuration is the possible wave propagation parallel to the optical axis in birefringent crystals such as KDP and $LiNbO_3$. Thus no compensation of temperature is needed and large apertures can be realized. The most important application of longitudinal modulators are Q-switches in high power lasers where nonactive losses have to be avoided. Therefore hole-electrodes are preferred to conducting transparent coatings. Some examples of electrode-designs are sketched in Fig. 5.3. The prominent deficiency of these cells is nonuniform transmission of the light beam across the effective aperture. Steinmetz et al.[115)] found that ring-electrodes produce a more homogeneous electric field than end-plate cells. Theoretical calculations, requiring the solution of at least a two dimensional Laplace's equation, have been carried out with the intention of optimizing the geometry of electrodes[116)]. Uniform fields and large apertures are achieveable by the use of plasma electrodes[117)].

A special type of a longitudinal device is the electrooptic Fabry-Perot modulator (Fig. 5.4). The reflected and transmitted intensities are given by[118)]

$$I_R = I_0 R \frac{4 \sin^2(\phi/2)}{(1 - R)^2 + 4 R \sin^2(\phi/2)}$$

$$I_T = I_0 \frac{(1 - R)^2}{(1 - R)^2 + 4 R \sin^2(\phi/2)}$$

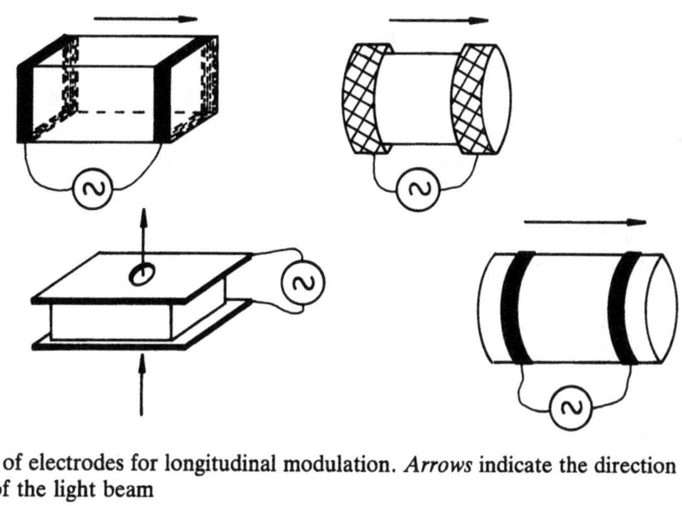

Fig. 5.3. Different shapes of electrodes for longitudinal modulation. *Arrows* indicate the direction of the effective field and of the light beam

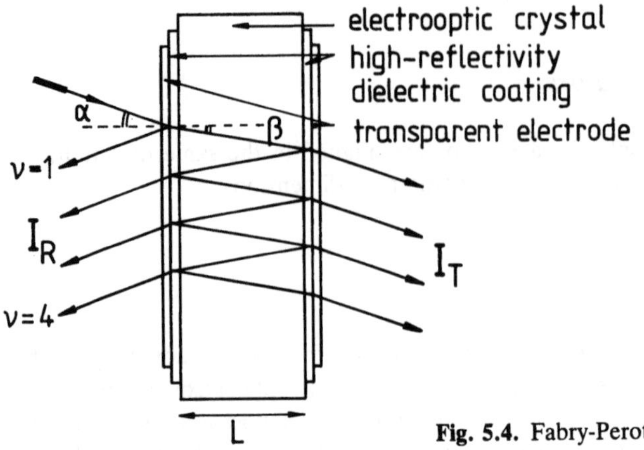

Fig. 5.4. Fabry-Perot modulator

where R denotes the reflectivity of the faces and $\phi = \frac{2\pi}{\lambda} 2nL \cos\beta$ measures the phase difference between two reflected beams I_R^ν and $I_R^{\nu+1}$. For $\phi = k \cdot 2\pi$ $(k = 0, 1, 2 \ldots)$ the transmitted intensity equals I_0.

If the Fabry-Perot resonator is biased so that $I_T = 0.5\, I_0$ in the absence of the modulator voltage, then

$$\partial I_T/\partial\phi \approx -I_0 \frac{\sqrt{R}}{1-R} = -I_0 \frac{F}{\pi}$$

for high values of R (>0.5). $F = \frac{\pi\sqrt{R}}{1-R}$ designates the finesse of the interferometer. In a simple modulator the maximal value of $\partial I/\partial\phi$ equals I_0. Thus the gain factor of a Fabry-Perot cavity is F/π, where a finesse of $F = 50$ is easily achieveable.

In a similar way improvements can be obtained with an electrooptical tuned Lyot-filter, consisting of several polarization modulators of different lengths[118)].

5.2.2 *Transversal Modulators*

In a transversal arrangement of the electric field and light propagation the half-wave voltage V_π can be reduced by increasing the sample length and decreasing the gap between both electrodes. As already discussed in Sect. 5.1.2 the improvement of efficiency by geometrical means is limited by diffraction effects. Another factor affecting the performance of a device is its sensitivity to thermal fluctuations. A modulator induces the phase difference

$$\phi = \frac{2\pi}{\lambda} \Delta n L$$

where Δn is $n'' - n'$ in a birefringent crystal or $n - 1$ in an interferometer as shown in Chap. 2. Assuming a sample length of 10 mm, a wavelength of 633 nm, a birefringence of $\Delta n = 0.1$, and a thermal coefficient of $dL/LdT = 10^{-4}\ K^{-1}$, a temperature change as small as 0.016 degrees induces an intensity variation of 1% of the full modulation depth. Temperature influences can be reduced by compensating samples. Compensation is also necessary when ray birefringence occurs.

A nice example of compensation technique is the ADP-four-crystal modulator recently reported by Moritani et al.[119]. The maximal electrooptic coefficient of ADP is r_{41} and therefore the optimal orientation is so as shown in Fig. 5.5. Ray-birefringence is compensated by sample II and IV and the tandems "I–II" and "III–IV" compensate the influence of temperature fluctuations. The device has been suggested as a modulator for polarization spectroscopy in ellipsiometry.

A transversal modulator manufactured from a quartz-crystal has recently been used for the measurement of high-voltage pulses to study the discharge kinetics of a XeCl laser[120]. Quartz-crystals have also been proposed for an electrogyration and Faraday-device, allowing a simultaneous measurement of electric field and currents[121].

An interesting application of the transversal configuration in an electrooptical Q-switch has recently be reported by Shaohuo[122]. He utilized the fact that for a light wave propagating nearly in the direction of an optical axis birefringence is only small and therefore can be influenced drastically by an external perturbation. He prepared an hexagonal shaped $LiNbO_3$-specimen with carefully orientated prism faces perpendicular to the (100) faces, which are supplied with electrodes. Ray-birefringence causes a spatial separation of both wave modes and total internal reflection allows the realization of different propagation directions. In this way the single-block hexagonal electrooptical Q-switch works as a modulator with integrated polarizer and analyzer.

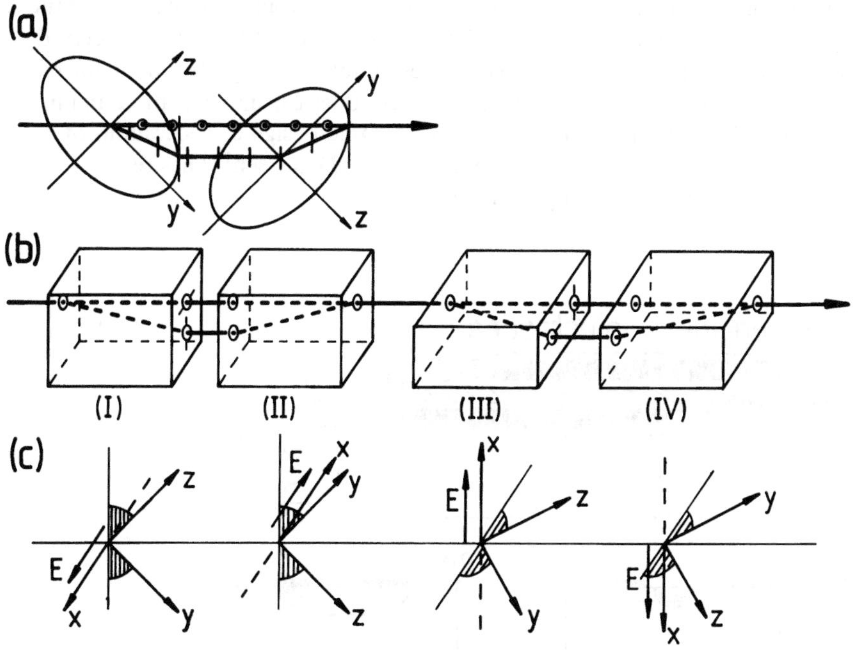

Fig. 5.5 a–c. Structure of a transversal ADP-modulator[119]. **a** Mechanism of compensation of ray-birefringence. **b** Beam propagation in the four ADP-samples. Sample *II* and *IV* compensate the ray-birefringence of sample *I* and *III*, respectively. Tandem *"III–IV"* compensates the wave-birefringence of tandem *"I–II"*. **c** Crystallographic orientation of the samples

5.3 Waveguides

5.3.1 Lumped Modulators

When a modulator is fabricated, people have two main aims: i) a small drive voltage and ii) a large bandwidth. Both conditions demand opposite dimensions of electrode-spacing and interaction length. Roughly spoken, bandwidth can be increased at the expense of drive voltage by simply increasing the device length. Thus, the appropriate figure of merit for a waveguide modulator is the bandwidth/voltage ratio $\Delta\nu/V$. To achieve an optimal performance of a device the relations considered in Sect. 5.1.2 and 5.1.3 are too simple and more exact calculations are necessary. For the capacitance limit $\Delta\nu/V$ is given by[123]:

$$\frac{\Delta\nu}{V} = \frac{n^3 \cdot r}{\pi \cdot R_T \cdot p \cdot \lambda \cdot \bar{\varepsilon}_{eff}} \frac{\Gamma}{G} K .$$

The material constants $n^3 \cdot r/\bar{\varepsilon}_{eff}$ indicate that crystals with high optical, but with a low dielectric response are desired. Therefore ferroelectric crystals may not be very suitable. R_T denotes the terminating resistance. Obviously the bandwidth can be increased by reducing R_T, but such an improvement has to be paid with an increasing drive power. The modulation depth is taken into account by p. For a 100% intensity change $p = 1$. G is the electrode gap and Γ is the overlap integral between the applied electric field and the electric field of the optical wave. Γ takes into account that in contrast to bulk devices both fields are inhomogeneous. The factor K stems from the calculation of the capacit C by conformal mapping techniques[123]. For a calculation of Γ and C the electrode structure has to be known. In Fig. 5.6 a typical design for a waveguide modulator on insulating crystals is illustrated. For such idealized electrodes calculations are found in references 124 and 113. Auracher and Keil[125] have reported design considerations of Mach-Zehnder interferometers and for this modulator type Danielsen[126] has recently carried out a two-dimensional propagating beam analysis.

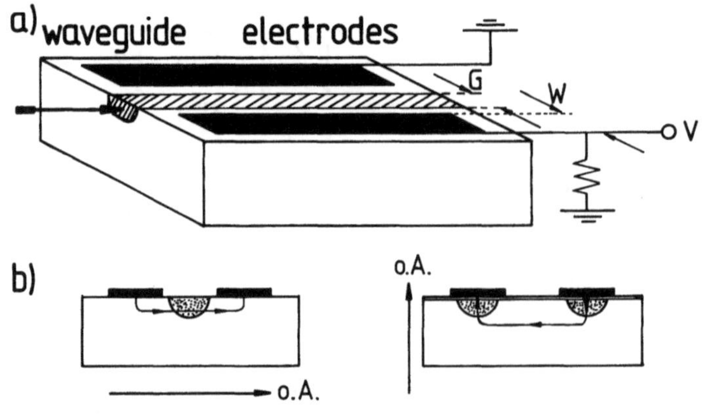

Fig. 5.6. a Typical design of an electrooptic waveguide modulator on the basis of $LiNbO_3$. **b** Arrangement of electrodes for utilizing horizontal or vertical electric fields. *o.A.:* optical axis

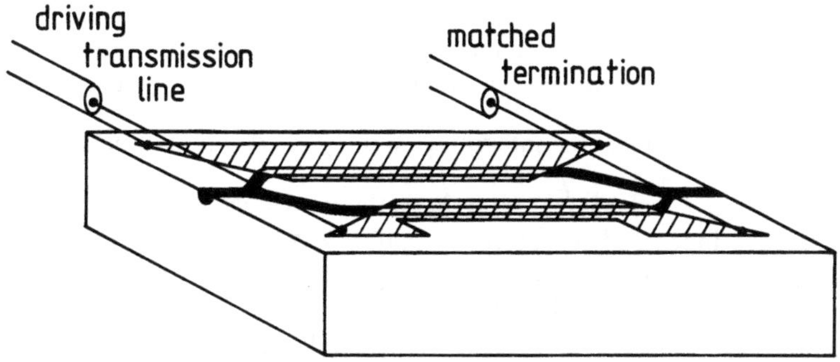

Fig. 5.7. An integrated y-branched Mach-Zehnder interferometer with traveling wave electrodes

The exact knowledge of theoretical models is sometimes not so important because significant improvements depend rather on the solution of technical problems during the fabrication process. This may be illustrated by some examples describing the temporal development of waveguide devices. The first lumped modulator has been demonstrated by Kaminow et al.[127] on the basis of $LiNbO_3$. The beam was guided in an outdiffused narrow ridge, confirming the light in both transverse directions. A phase modulation of π, corresponding to an on-off-switch in an intensity modulator, was achieved with 3.8 V and the bandwidth limit was 640 MHz. One year later the same group[128] presented a Ti-diffused $LiNbO_3$ strip-waveguide, which reduced the required power by more than 10%. Neyer and Sohler[129] demonstrated a cutoff modulator with more than 1 GHz bandwidth. In a $LiNbO_3$ crystal they prepared by Ti-diffusion a 2 μm wide and 3.2 mm long channel in such a way that the condition for guiding a 633 nm-ray was just fulfilled for the TE_{00} mode. If the index of refraction in the channel is lowered by an electrical field, the light is forced to couple out into the substrate. In this way a 70% intensity change was obtained with a voltage of 11 V. Improvements in preparation techniques allowed the production of a more complicated structure such as an y-branched Mach-Zehnder interferometer, allowing the application of push-pull electrodes. A typical example of a Mach-Zehnder interferometer is shown in Fig. 5.7, but with traveling wave instead of lumped electrodes.

3 dB-bandwidth of 1.1 GHz[130] and 1.4 GHz[131] has been achieved. Using high-resolution fabrication techniques it was possible to prepare electrodes with only 1 μm spacing, enabling a short device (750 μm) which again increased the bandwidth to 4.5 GHz[132]. Minimizing electrode resistance and inductance it was even possible to obtain a 3 dB-bandwidth of 7.3 GHz[133]. Wong et al.[134] demonstrated a $LiNbO_3$ modulator fabricated by proton exchange in benzoic acid. They found its performance comparable with that of a Ti-indiffused device. Its advantage is the simplicity in fabrication. Recently Yajima et al.[135] manufactured an optical-optical waveguide modulator, which obtains the drive voltage from an integrated silicon photovoltaic detector.

5.3.2 Traveling Wave Modulators

In a lumped modulator the bandwidth is limited by the electrode charging time. This limit is overcome if the characteristic impedance of the electrodes is equalized to that of the cable and they are connected as shown in Fig. 5.7. Now the device is integrated in the driving transmission line and the important time factor is the difference in transit time for the optical and RF waves. The induced optical phase shift is no longer homogeneous within the sample. Its integrated value $\int_0^L \Delta\phi \, dz$ is proportional to $\sin x/x$, where $x = \pi(N_m L_m - nL)/\lambda_m$[123]. Here λ_m is the wavelength of the microwave, $N_m = \sqrt{\varepsilon_{eff}}$ its refractive index and n that of the optical wave. $\sin x/x = 1$ for $x = 0$. Its value drops rapidly for increasing x. The condition $(N_m L_m - nL) = 0$ can be realized in two ways. As $N_m > n$, the light wave is faster than the microwave, but the microwave velocity v_m can be increased by filling a transmission line only partially with the crystal. Then the effective v_m is a mixture between the value in the sample and in air[110]. However, this kind of velocity matching requires high voltages and is not convenient for waveguides. The second way is to direct the light beam obliquely with respect to the wave vector of the microwave field. The conditions for the conservation of energy and momentum have been analyzed by Rigrod and Kaminow[136] for a KDP-crystal. This kind of matching leads to zig-zag-modulators which have been realized mainly in bulk devices[111].

In waveguides $\vec{k}_{light}$ is usually parallel to $\vec{k}_{microwave}$ and there is a velocity difference between the two waves. For a $LiNbO_3$-rod of 1 cm length the bandwidth limit due to this mismatch is about 7 GHz. It is essentially higher (~5 times) in GaAs because N_m is smaller and n larger in the semiconductor than in lithium niobate[123]. Another effect, which reduces the signal is the microwave loss by the electrodes. Kubota et al.[137] have studied the electrode design of a lithium niobate traveling wave modulator theoretically and experimentally. They found, that asymmetric electrodes are superior to symmetric ones, that the thickness of Al-electrodes should be at least 3 μm, and that with an increasing width to gap ratio the electrode conductor losses decrease. Experimentally a loss of 1 $dB/cmGHz^{1/2}$ was achieved, which reduces the signal by a factor of 2 at 10 GHz, when the electrodes are 1.5 cm long.

An excellent review about the development of waveguide modulators until 1982 is the already cited report of Alferness[123]. More recently Gee et al.[138] described a typical Mach-Zehnder interferometer with traveling wave electrodes. Its design is illustrated in Fig. 5.7. They achieved a constant phase bias of $\pi/2$ or π by introducing slightly different lengths of both arms of the interferometer. With a bias of $\pi/2$ the device functions as an intensity modulator and with a bias of π as a frequency shifter. For utilizing the large coefficient r_{33} in a z-cut $LiNbO_3$ substratum the electrodes have to be placed directly on top of the two waveguide sections. To avoid optical losses by the electrodes a 250 nm SiO_2 buffer layer is sputter-deposited on the substratum. The electrode gap is determined by the waveguide separation and the length is restricted to 6 mm for avoiding problems with velocity mismatch. Now the width of the electrodes can be prepared so that their characteristic impedance is 50 Ω or that their resistivity losses are minimal. The chosen compromise results in a 3 dB bandwidth of 3.5 GHz limited by microwave losses. Complete modulation is achieved with 6.5 V at 0.83 μm wavelength and with 18 V at 1.3 μm wavelength. Thus the increase of drive voltage, which is necessary when the wavelength is changed from the visible to the infrared region, is much greater than a linear increase.

Recently Becker[139] has demonstrated that it is possible to achieve the theoretical bandwidth determined by the velocity difference between the microwave and the optical waves. With a Mach-Zehnder interferometer he obtained a 3-dB bandwidth limit of 16 GHz for a 4 mm long interaction length. At 850 nm the halfwave voltage was

$V_\pi = 4.5$ V. The improvement in comparison to earlier devices he achieved mainly by minimizing the dc resistance and the surface roughness of the electrodes.

Up to now we have described only experiments which test the frequency response of a device. However, the availability of picosecond and subpicosecond laser pulses has made investigations in the sphere of time possible. Izutsu and Sueta[140] have analysed theoretically and experimentally the time response of phase and intensity modulators of the traveling-wave- type. Valdmanis et al.[141] have used 100-fs pulses from a colliding pulse mode-locked system to generate the electrical signal via a Cr : GaAs photoconductor. A second beam of pulses probed the birefringence induced by the electrical pulses as they propagate down a transmission line built on $LiTaO_3$. A temporal resolution of less than 500 fs has been achieved. This high speed was possible because the probe beam entered the crystal with such an angle of incidence that the velocity of the optical and electrical signals were matched[142].

5.3.3 Electrooptical Mode Coupling

As shown in Chap. 2.5 in a waveguide several modes with different propagation constants β may exist. Combining Eq. (2.8) and (2.11) the relation between β and the thickness x_0 of the guiding layer is found to be

$$\frac{x_0}{\lambda} = \frac{1}{2\,\pi[n_2^2 - \beta^2/k_0^2]^{1/2}} \left\{ m \cdot \pi + \tan^{-1}\left[\frac{\beta^2/k_0^2 - n_1^2}{n_2^2 - \beta^2/k_0^2}\right]^{1/2} + \tan^{-1}\left[\frac{\beta^2/k_0^2 - n_3^2}{n_2^2 - \beta^2/k_0^2}\right]^{1/2} \right\}$$

for a TE_m mode and a similar expression can be derived for a TM_m mode. In electrooptic modulators the region I (Fig. 2.3) is usually air and $n_1 = 1$. For this condition Fig. 5.8 shows the dispersion of the modes in a $LiNbO_3$-waveguide. If the thickness to wavelength ratio x_0/λ is so small that $\beta/k_0 = n_3$, there is no optical boundary between the guiding region II and the region III. Below this cutoff-value the mode is no longer confined. If x_0/λ is chosen in such a way, that β/k_0 is only slightly larger than n_3, small changes of the refractive indices by an electric field may change a confined mode into a radiation mode. This effect has already been used in an electrooptical device[129]. There exist even more mechanisms which are typical for waveguides and which cannot be observed in bulk crystals. The most important one in respect of electrooptic modulation is the coupling of two modes. The theory of mode coupling in optical waveguides has been described by Yariv[143] and Marcuse[144]. Here two cases are of interest, firstly the coupling of a TE- and a TM-mode and secondly the coupling of two waves which propagate in the same direction and are separated by only a small distance. The devices based on both effects are described in the two next chapters.

In general the coupling of two normal modes ξ_A and ξ_B with amplitudes A(z) and B(z) are described by[8, 143, 144]

$$\begin{aligned} \frac{d}{dz} A(z) &= -i\varkappa B\, e^{i\Delta\beta z} \\ \frac{d}{dz} B(z) &= -i\varkappa A\, e^{-i\Delta\beta z} \end{aligned} \tag{5.5}$$

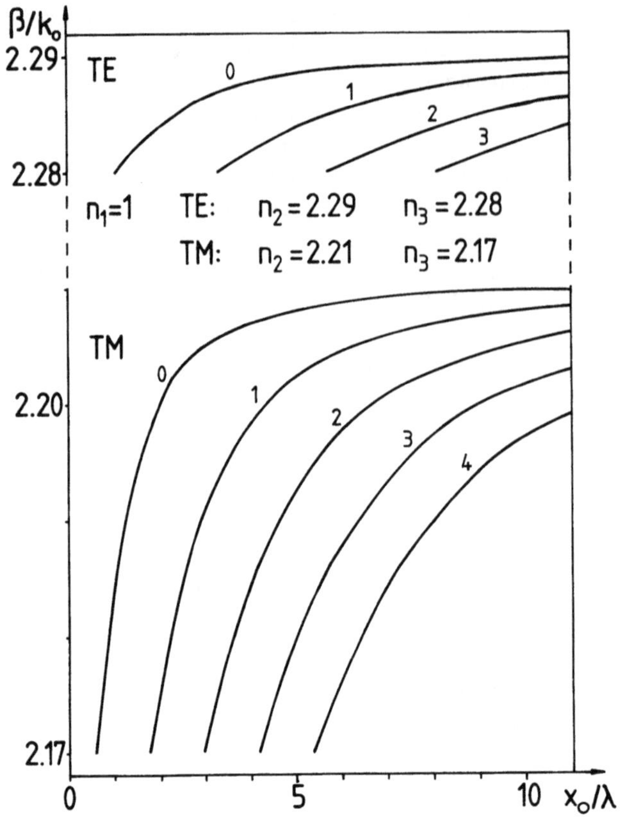

Fig. 5.8. Dispersion curves for confined modes in a $LiNbO_3$: Ti waveguide

where $\varkappa$ is the coupling coefficient and $\Delta\beta$ is the difference of both propagation constants. Assuming that at $z = 0$ $A(0) = A_0$ and $B(0) = 0$, the solution of Eq. (5.5) is given by[8)]

$$A(z) = A_0 e^{i\delta z}\left\{\cos[(\varkappa^2 + \delta^2)^{1/2} z] - i\frac{\delta}{(\varkappa^2 + \delta^2)^{1/2}} \sin[(\varkappa^2 + \delta^2)^{1/2} z]\right\}$$

$$B(z) = -iA_0 e^{-i\delta z} \frac{\varkappa}{(\varkappa^2 + \delta^2)^{1/2}} \sin[(\varkappa^2 + \delta^2)^{1/2} z] .$$

Here the abbreviation $2\delta = \Delta\beta$ has been used. At $z = L$ the fraction of power which is transferred from guide A to guide B equals

$$\eta = \frac{BB^*}{AA^* + BB^*} = \frac{\varkappa^2}{\varkappa^2 + \delta^2} \sin^2[(\varkappa^2 + \delta^2)^{1/2} L] . \tag{5.6}$$

Complete power transfer occurs if $\delta = 0$ and $L = \pi/2\varkappa$. For $\delta = 0$ the phase velocities in the two modes are equal (phase matching). In electrooptic devices the coupling constant $\varkappa$ and the phase difference δ are changed by the modulation field.

5.3.4 TM ↔ TE Conversion

If the coefficient $\varkappa$ in Eq. (5.5) couples a TE-mode with a TM-mode, it is possible to convert the polarization of a light wave. A necessary condition is the induction of a dielectric polarization $\vec{P}(\omega)$ perpendicular to the field $\vec{E}(\omega)$ of the incident wave. This condition can be fulfilled by an externally applied electrical field E_k if an electrooptic coefficient r_{ijk} with $i \neq j$ exists. As shown in Table 4.1, in $LiNbO_3$ two coefficients of this type exist: $r_{61} = -r_{22}$ and r_{51}. With the modulation field parallel to the crystallographic a-axis and the light wave propagating along the c-axis, electrooptically induced conversion occurs via the r_{61} coefficient. The advantage of this configuration is the orientation of the beam parallel to the optical axis. As natural birefringence is absent, phase matching can easily be realized. Thus polarization conversion is independent of the operating wavelength[145)]. Unfortunately the magnitude of r_{61} is small in $LiNbO_3$. As shown in Table 4.4, r_{51} is five times larger. For taking advantage of this coefficient the light beam has to propagate along b* (perpendicular to a and c) and the modulation field is orientated as before. Now the phase velocities of both modes are significantly different because the natural birefingence is large. Nevertheless phase matching can be achieved by electrooptically induced corrugations. They are obtained by supplying a waveguide with interdigital electrodes as illustrated in Fig. 5.9. Let us assume that the light wave is traveling parallel to z and the $LiNbO_3$ sample is orientated in such a way that z coincides with b* and x with the a-axis. Then the field component parallel to x in Fig. 5.9 induces a periodically varying coupling coefficient

$$\varkappa = \varkappa_0(E)\, e^{-i2\pi/\Lambda} \tag{5.7}$$

via the electrooptic coefficient r_{51}. Here we consider only one component of the Fourier series which describes the periodicity of $\varkappa$ more completely. Combining the space varying part in Eq. (5.7) with that in Eq. (5.5) the phase angle $\Delta\beta$ is given by

$$\Delta\beta = \beta_A - \beta_B - 2\,\pi/\Lambda = 2\,\delta\,.$$

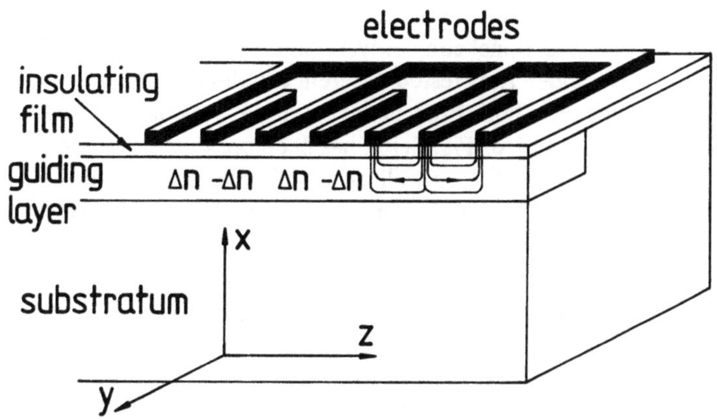

Fig. 5.9. Interdigital electrodes on a Ti:$LiNbO_3$-waveguide. In the space between the electrodes the z-component of the field is large, but under each electrode the field is mainly parallel to x. The effect of each field component depends upon the orientation of the crystal

Now the condition for phase matching ($\delta = 0$) reads

$$\beta_{TE} - \beta_{TM} = \frac{2\pi}{\lambda_0}(N_{TE} - N_{TM}) = \frac{2\pi}{\Lambda}. \tag{5.8}$$

With the numerical values of Fig. 5.8 the electrode period Λ has to be 7 μm for $\lambda = 0.6$ μm.

The first $LiNbO_3$ mode converter has been demonstrated by Alferness[146]. Optimizing the design of the electrodes complete TE ↔ TM conversion has been achieved with as little as a 2.5-V drive[147]. This value should be compared with the 52 V required with the above mentioned device which utilizes the coefficient r_{61}[145]. The modulation bandwidth and efficiency which have been achieved with an electrooptic polarization modulator are similar to those of other lumped modulators. Due to the phase matching condition (5.8) a device with periodic electrodes is strongly wavelength selective, allowing special applications such as wavelength multiplexing[148]. On the other side wavelength tuning with periodic electrodes is also possible by supplying the waveguide with a second system of electrodes. In the nomenclature of Fig. 5.9 this second system produces a field parallel to y, which is the crystallographic c-axis. Now the electrooptic coefficients r_{33} and r_{13} modify N_{TE} and N_{TM} in Eq. (5.8) and thereby the phase matching wavelength is changed[149]. Due to the coefficient r_{41} in cubic crystals of symmetry $\bar{4}3m$ modulation by TE ↔ TM conversion without periodic electrodes is possible with semiconductors, too[150].

5.3.5 *Directional Coupling*

At the boundary between the guiding layer and the substratum the field of a confined mode does not stop abruptly but escapes exponentially. If two waveguides are close together as illustrated in Fig. 5.10 their evanescent fields overlap and the mutual perturbation causes a coupling. The coupling coefficient $\varkappa$ depend upon the waveguide parameters, wavelength, and interwaveguide separation[151]. Whereas in the polarization conversion of the last chapter the electric field $\vec{E}$ produces the coupling, in directional coupler switches $\vec{E}$ modulates the phase difference δ.

The first directional coupler was demonstrated by Campbell et al.[152] on the basis of a GaAs-waveguide. They tested different coupling lengths and guide separations. A typical distance between both branches is 3 μm. A popular $LiNbO_3$-modulator with the electrode configuration of Fig. 5.10a has been presented by Papuchou et al.[153]. They called it “Cobra”. Lumped[154] as well as traveling wave electrodes have been applied. In the latter case a 3 dB modulation bandwidth of 7.2 GHz for an optical wavelength of 1.32 μm has been achieved[155].

As already discussed in Chap. 5.3.3 maximal power transfer between two modes occurs for a special length, the single coupling length $L_0 = \pi/2\varkappa$. Stringent fabrication tolerances exist to meet the requirement that L equals L_0 or is an odd multiple of L_0.

Furthermore electrode gap and guide separation are not independent of each other, but each quantity has to be optimized in respect of different property: in respect of low capacitance and optical coupling, respectively. Different tricks have been tried to facilitate the experimental situation. Mikami and Noda[156] proposed to load a metal film on one of the waveguides to influence the effective coupling length. The most important method has been invented by Schmidt and Kogelnik[157]. They applied reversed $\Delta\beta$-electrodes as shown in Fig. 5.10b. Devices with six sections of alternating $\Delta\beta$ have been realized[158]. One advantage of a reversed $\Delta\beta$-coupler is the small total value of δ in the factor $\varkappa/(\varkappa^2 + \delta^2)^{1/2}$ in Eq. (5.6). The coupling coefficient can be determined in the straight-through state or by measuring the efficiency of second harmonic generation as suggested by Bozhevol'nyi[159]. Alferness[160] has suggested a weighted coupling by varying the distance of the electrodes as illustrated in Fig. 5.10c. Combining this method with reversed $\Delta\beta$-electrodes a modulation which is independent of light polarization is possible[161]. This property is important in applications with glass fibres which do not preserve the polarization state of light. A recent theoretical analysis of coupling length takes into account the effects of corner regions of a rectangular waveguide[162]. Directional coupling is not restricted to two guides. In a theoretical analysis the principle of alternating $\Delta\beta$ has been applied to a system of multiple coupled waveguides[163]. A three-guide coupler in the input and output section of an interferometer has been used to fabricate a GaAs modulator[164]. The bandwidth was 2.2 GHz which is a small value compared with $LiNbO_3$ devices. Another deficiency of semiconductors is their large optical loss. A compensation of this loss by combining directional coupler switching with optical amplification by laser action has recently been discussed[165].

A completely different mechanism of directional coupling in an y-junction waveguide is illustrated in Fig. 5.10d. The electric field changes the refractive index in both branches with an opposite sign because of the opposite field directions. Therefore the conditions for waveguiding are different in both branches. Deflection angles of $1°$[166] to $2°$[167] are possible.

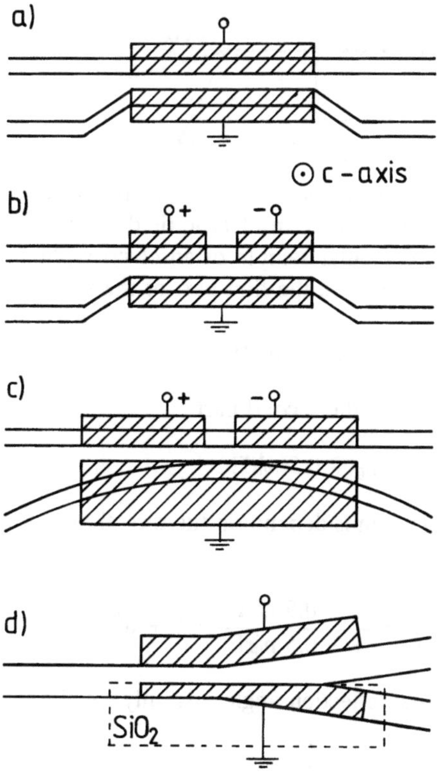

Fig. 5.10 a–d. Directional coupling in Ti : $LiNbO_3$-waveguides. **a** Cobra configuration, **b** reversed $\Delta\beta$ coupler, **c** weighted coupling with reversed $\Delta\beta$ electrodes, **d** Y-branch mode splitter

A similar type of waveguide junctions is the crossing-channel waveguide. It is often assumed to operate via the mechanism of total internal reflection (TIR). TIR-devices are described in Chap. 5.4.4. However, theoretical considerations support the assumption of mode interaction[168)] or divide these modulators into two classes, reflective-type and mode interference-type devices[169)]. The distinction between both types depends on various parameters, such as the optical wavelength, the waveguide width, the waveguide crossing angle, and the number of modes supported by the waveguide.

5.4 Deflectors

5.4.1 Bulk Prism Deflectors

The process of light deflection can easily be demonstrated with a simple prism. Referring to the geometry of Fig. 5.11 the application of Snell's law yields

$$\frac{n_1}{n_2} = \frac{\sin(\theta + \alpha)}{\sin\alpha} = \sin\theta \cdot \cot\alpha + \cos\theta \,. \tag{5.9}$$

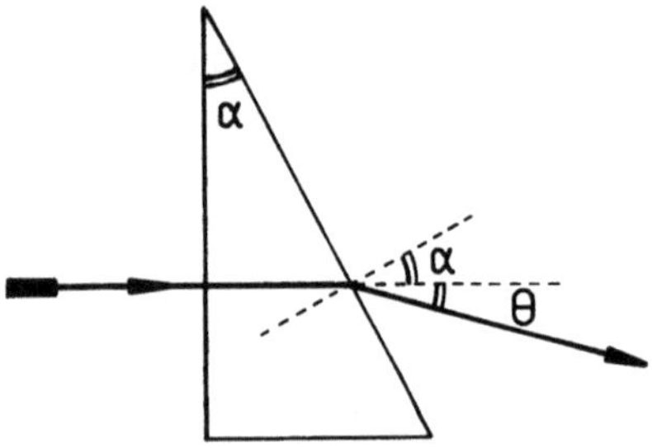

Fig. 5.11. Deflection of light by a prism

In an electrooptic device the difference in the indices of refraction, $n_2 - n_1 = \Delta n$, is produced by an electric field. A possible arrangement is shown in Fig. 5.12a, where a rectangular sample of an electrooptic crystal is supplied with two separated electrode-systems. In the first part of the specimen the index of refraction is

$$n_1 = n_0 \pm 0.5\, n_0^3 rE$$

and in the second part

$$n_2 = n_0 \mp 0.5\, n_0^3 rE \,.$$

The signs depend on the direction of $\vec{E}$ and on the sign of the electrooptic coefficient r. As

$$\frac{n_1}{n_2} = \frac{1 + 0.5\, n_0^2 rE}{1 - 0.5\, n_0^2 rE} \approx 1 + n_0^2 rE \approx 1 \,,$$

the deflection angle θ is only small and Eq. (5.9) reduces to

$$\theta \cot\alpha = n_0^2 rE \,.$$

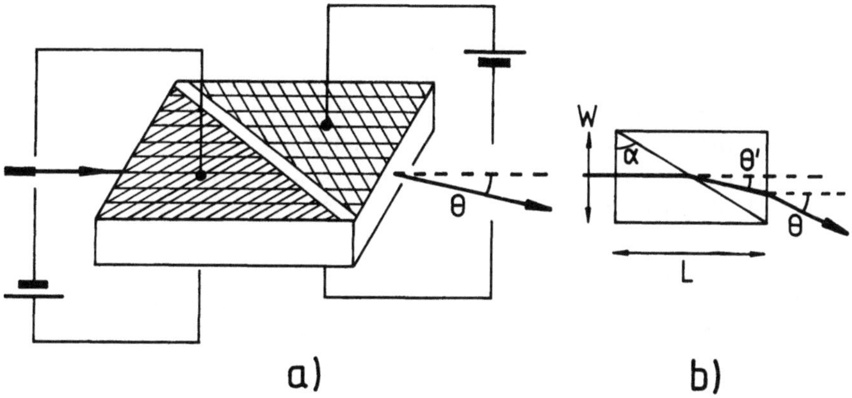

Fig. 5.12 a, b. Electrooptic prism deflector

Then θ′ in Fig. 5.12b is given by

$$\theta' = \tan\alpha \; n_0^2 rE = L n_0^2 rE/W \; .$$

Due to the refraction at the rear face

$$\theta = \theta' \, n_0 = L n_0^3 rE/W \; .$$

The magnitude of θ is not the critical point for an efficient deflection, because θ can easily be magnified with a passive element like a lense. More important is the spatial resolution of the light spots. Due to diffraction of light at the aperture W, the beam becomes divergent. From diffraction theory the half angle divergence is given by

$$\delta\theta = \frac{\lambda}{W} \; .$$

The Rayleigh criterion says, that two spots are just resolved, when the angular displacement θ of the beam equals δθ. Then the relation

$$N = \frac{\theta}{\delta\theta} = \frac{L}{\lambda} n_0^3 rE \tag{5.10}$$

determines the number of spots, which can be resolved. If deflection is performed in both directions, which can be realized by reversing the field, N becomes twice the value of (5.10).

In the arrangement of Fig. 5.12a the opposite piezoelectric deformation in both parts causes mechanical stresses in the boundary face which distort the beam and reduce the lifetime of the device. Therefore two separated triangular samples are more convenient. Both pieces are inversely orientated and common electrodes are applied, preventing fringing of the electric field from one prism to the other. Both prisms have to be joined with a matching index material to eliminate reflection losses at the interface. Theoretical descriptions of bulk deflectors are found in reference 170.

Haas et al.[171)] have used a $BaTiO_3$-prism with an apex angle of 20°59′ and a thickness of 0,73 mm in the minimum deviation position. The sample was heated 3–7 °C above the paraelectric-ferroelectric phase transition and consequently they observed a quadratic dependence of the deflection on the applied field. Applying an electrical field of about 5000 V/cm they achieved a deflection of 3.3 min of arc.

A prism deflector of higher efficiency has been presented by Chen et al.[172)] who fabricated a device from the $KTaO_3/KNbO_3$ mixed crystal KTN, which has a T_c of 283 K, just below room temperature. With a prism of a base length of 3.8 mm, a thickness of 1.6 mm, and an angle of 46° they achieved an angular deflection of 34 min of arc. 8 spots of a HeNe laser could be resolved completely, which corresponds to 15 resolved spots based on the Raleigh criterion. Due to inhomogeneities of the material the full width of the prism could not be used and diffraction effects produced a relatively large divergence of the beam of 10.7 min of arc. For an optimal KTN-device they estimated $N = 1.2 \times 10^4$. With such a deflector it is possible to project a two-dimensional picture with a drive power of 37 mW/MHz.

Kerr-effect deflectors are only convenient when advantage is taken of the critical divergence of the dielectric constant $\tilde{\varepsilon}$ near a phase transition. A large $\tilde{\varepsilon}$ is connected with a high electrooptic response and reduces fringing effects of the electric field. Furthermore the symmetry which is usually cubic above T_c allows a larger aperture than in birefringent crystals. On the other hand the critical divergence of $\tilde{\varepsilon}$ requires a good temperature stabilization and is related to a large electrocaloric effect, which is proportional to $\partial\tilde{\varepsilon}/\partial T$ and warms up the sample.

5.4.2 Gradient-Type Deflectors

Deflection does not only occur at a wedge-shaped interface but also in a material with a space varying index of refraction. Consider the rectangular sample in Fig. 5.13 which should possess a refractive index with the linear increase $\partial n/\partial x_2$. Then the two beams A and B traverse the medium with the velocities $v_A = \dfrac{\nu \cdot \lambda}{n}$ and $v_B = \dfrac{\nu \cdot \lambda}{n + \dfrac{\partial n}{\partial x_2} W}$.

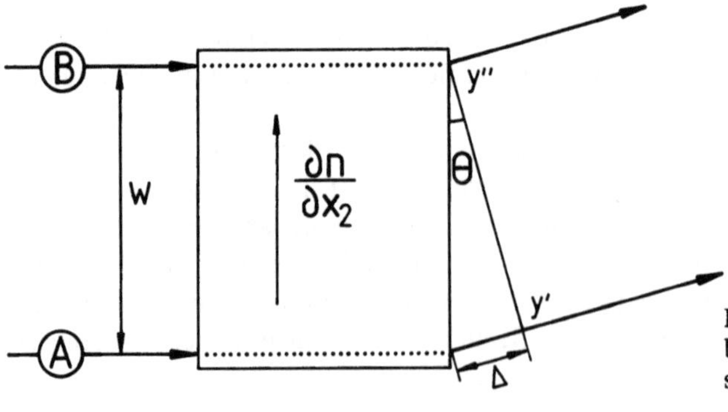

Fig. 5.13. Deflection of a light beam by a medium possessing a space varying index of refraction

At time t their common wave front is y′y″ and the distances traveled are

$$S_B = L = v_B t \quad \text{and}$$

$$S_A = L + \Delta = v_A t' + v_0(t - t') = L + v_0 t - \frac{v_0}{v_A} L ,$$

v_0 being the velocity in vacuum. Eliminating the propagation time t, the spatial excess Δ becomes

$$\Delta = \left(\frac{v_0}{v_B} - \frac{v_0}{v_A}\right) L = \frac{\partial n}{\partial x_2} WL$$

and the deflection angle is

$$\theta \approx \sin\theta = \frac{\Delta}{W} = \frac{\partial n}{\partial x_2} L . \tag{5.11}$$

The gradient of n is achieved by supplying a sample with nonplanar electrodes, producing for instance a quadrupole field. In Fig. 5.14 an idealized design with hyperbolic electrodes is shown[173)]. Let us assume that for a light beam propagating along x_3 the only effective electrooptic coefficient is r_{23}^{eff}. This is the case in a KDP-crystal with x_3 parallel to the optical axis and $r_{23}^{eff} = 2\, r_{63}$. The refractive index varies as

$$\frac{\partial n_2}{\partial x_2} = -0.5\ n_2^3 r_{23}^{eff} \frac{\partial E_3}{\partial x_2} \tag{5.12}$$

and the potential V in the deflector is described by

$$V(x_2,\ x_3) = \frac{2\,V_0}{R_0^2}\, x_2 x_3 ,$$

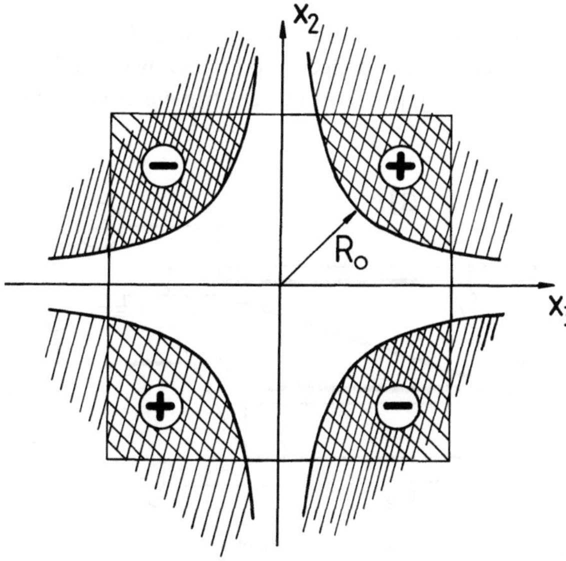

Fig. 5.14. Hyperbolic electrodes of a quadrupole deflector

which results in the constant value

$$\frac{\partial E_3}{\partial x_2} = \frac{2\,V_0}{R_0^2}\,. \tag{5.13}$$

Combining Eqs. (5.11)–(5.13) we get for the deflection angle

$$\theta = n^3 r L \frac{V_0}{R_0^2}\,.$$

Then the number of spots, which can be resolved, is given by

$$N' = \frac{W}{\lambda}\,\frac{L}{R_0^2}\,n^3 r V_0\,.$$

The last expression should be compared with Eq. (5.10) for a simple prism deflector:

$$N'' = \frac{1}{d}\,\frac{L}{\lambda}\,n^3 r V_0\,,$$

where d designates the thickness of the prism. There is no fundamental advantage of N′ over N″, which can readily be seen by putting $W \approx R_0 \approx d$. But the gradient-type deflector is superior to the prism in respect to less distortions of the wave front[174]. On the other hand ideal hyperbolic electrodes cannot be realized and some theoretical efforts have been undertaken to determine a design which preserves constant gradients and can easily be manufactured[175].

5.4.3 *Surface Prism Deflectors*

Nearly at the same time Kaminow and Stulz[176] and Tsai and Saunier[177] have transferred the ideas of a prism deflector to $LiNbO_3$ waveguides. Their devices consist of a rectangular specimen supplied with three strip electrodes which simulate a prism structure. With

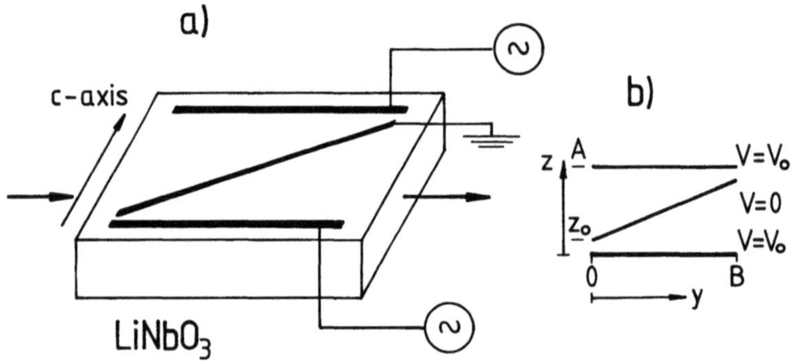

Fig. 5.15 a, b. Surface prism deflector

the orientations of Fig. 5.15 the actual electrooptic coefficient is approximately r_{33} and the total field induced change of refractive index n_e is expressed as

$$\Delta n_e = n_e^3 r_{33} \int_0^B E_2(z, y)dy .$$

Obviously the properties of the surface prism deflector are intermediate between those of a simple prism- and a gradient-type-deflector. The problem in determining Δn_e is the calculation of the electric field. Two different approximations have been reported for the number of resolvable spots. Assuming that the width of the diagonal electrode, the thickness of the planar waveguide, and z_0 are much smaller than the spacing A, Kaminow and Stulz[176)] derived the expression

$$N = \frac{n_e^3 r_{33} BD}{\lambda A^2} V_0 .$$

They considered a Gaussian beam of diameter D at its waist and applied the $1/e^2$-criterion, which states that two spots are resolved when the intensity between them is dropped to $1/e^2$. The geometrical data of their device are $A = 100\,\mu m$, $B = 6\,mm$, $z_0 = 10\mu m$, and $D = 6.6\,\mu m$. Experimentally they resolved two spots with $V_0 = 30$ V.

The expression reported by Tsai and Saunier[178)] reads

$$N = \frac{n_e^3 r_{33} B}{\lambda(A - z_0)} \ln\left(\frac{A}{z_0}\right) V_0 .$$

With the dimensions $z_0 = 9$ µm, $A = 150$ µm, and $L = 10$ mm, they needed $V_0 = 8$ V for complete resolution of two spots of a He-Ne-laser. Mounting four single deflectors in a parallel array the capacitance was about 4.0 pF, allowing subnanosecond switching-speeds. With a 50 Ω termination the possible bandwidth of the device is 1.6 GHz, which corresponds to a drive power of 0.8 mW/MHz for 100% intensity modulation. This value should be compared with the 37 mW/MHz of the KTN bulk deflector described earlier.

Several theoretical efforts have been made to analyze the electrooptic surface prism waveguide deflector more thoroughly[179)]. Lee et al.[180)] tried different shapes of the central electrode and improved thereby the quality of the diffraction spots.

5.4.4 Total Internal Reflection and Bragg Deflectors

Comparing bulk deflectors, including quadrupole- and prism-types, with waveguide deflectors, there are different advantages and disadvantages of both kinds. Waveguide devices work with low drive voltage and they need electrodes on only one surface but the aperture is limited and the tolerances in coupling the optical beam are small. Roughly speaking, the arguments are reversed for bulk devices. Chang[181)] has tried to combine the advantages of both types by using the effect of total internal reflection (TIR). The application of TIR in integrated optics has been demonstrated by Tsai et al.[182)]. Using a waveguide structure with two crossed channels the outcoming light can be switched between two directions. The deflection angle is not substantially larger than in surface prism deflectors.

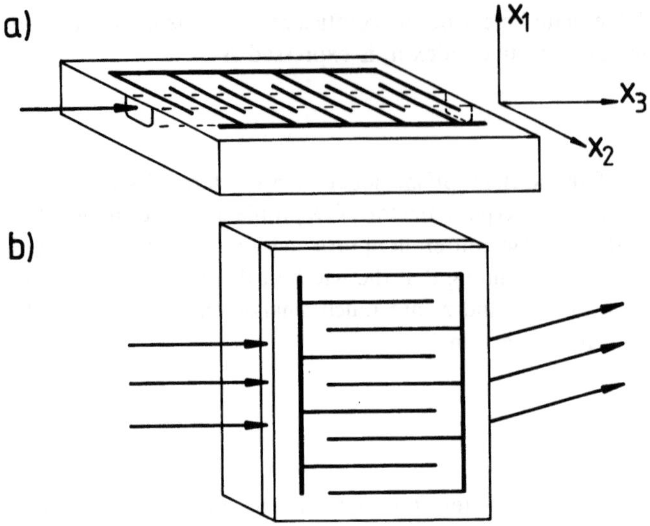

Fig. 5.16 a, b. Different effects and orientations in electrooptic modulation by waveguides. **a** TE ↔ TM-mode Conversion. The reference system indicates the crystallographic orientation of a $LiNbO_3$-sample. The vertical component of the electric field and the coefficient r_{51} are utilized. **b** Configuration for Bragg reflection

In Chap. 5.3.4 we have already shown that the periodical modulation of the refractive index may produce a coupling between two waves in a waveguide. The coupling occurs when the wavevector is orientated relative to the grating as illustrated in Fig. 5.16a. In orientation b) of the same figure Bragg reflection occurs. The deflection angle $2\,\theta_B$ is given by

$$\sin\theta_B = \lambda_0/2\,\Lambda N$$

where Λ is the period of the interdigital electrodes and N the effective refractive index of the guided wave.

The first electrooptic Bragg deflector has been demonstrated by Hammer and Phillips[183)] with a $LiNb_xTa_{1-x}O_3$-waveguide on a $LiTaO_3$-substratum. They provided the (010)-plane of the crystal with the grating and used for the direction of the interdigital electrodes an angle ϕ relative to the c-axis. The magnitude of ϕ was chosen so that the effective electrooptic coefficient and waveguiding, which depends on the index difference between the niobate and the tantalate, were both optimized. A theoretical analysis of a rectangular Bragg deflector has been reported by Belin et al.[184)]. They obtained a rather complicated expression for the figure of merit of a Bragg deflector. A more simple relation is

$$Q = 2\,\pi\lambda_0 L/N\Lambda^2\,.$$

Bragg reflection occurs most efficiently when $Q > 10$[185)]. In integrated switching networks the crossing section of two waveguide channels are provided with grating electrodes. Then light can be switched from one channel into the other one by Bragg reflection[186)].

5.5 Miscellaneous Devices

5.5.1 Electroabsorption

As discussed in Chap. 4.5.5 with KDP-crystals the lowest lying absorption band shows sometimes an only small electrooptic response. Then the use of light with wavelengths close to the absorption edge gives no advantage because optical losses are not compensated by a resonance enhancement. However, the situation is different in semiconductors with an directly allowed electric dipole transition from the valence to the conduction band. Such a material is GaAs. The dependence of its absorption coefficient α on the externally applied electric field is illustrated in Fig. 5.17. Obviously the expansion of α as a power series (Eq. 2.5) is no longer an appropriate description of electroabsorption. In GaAs[187], in InP[188], and in the quarternary alloy $Ga_{0.24}In_{0.76}As_{0.52}P_{0.48}$[189] it was possible to describe the electroabsorption effect in terms of a microscopic model, known as the "Franz-Keldysh-effect". This effect causes also a strong electrorefraction[188] and a strong Kerr effect for photon energies near the band gap[191]. Maksimov et al.[192] have compared the efficiency of electroabsorption and of the linear electrooptic effect in GaAs. At a wavelength of 0.902 μm, where optical losses are only small, they determined for a 90% modulation a necessary length of 0.16 mm for an electroabsorption device and a length of 1.6 mm for a Pockels cell.

The application of electroabsorption and electrorefraction in modulators is not the only way to utilize the Franz-Keldysh-effect. As shown in Fig. 5.18 for a photon energy slightly smaller than E_g a photocurrent is obtained. It has been used to fabricate avalanche photodiodes based on GaAs[193]. An interesting application of the Franz-Keldysh-effect in a GaAs charge-coupled device (CCD) has been reported by Kingston et al.[194]. In a CCD charges are moved in the surface region of a semiconductor sample by varying the voltage applied to strip electrodes[195]. CCD's based on GaAs have principal

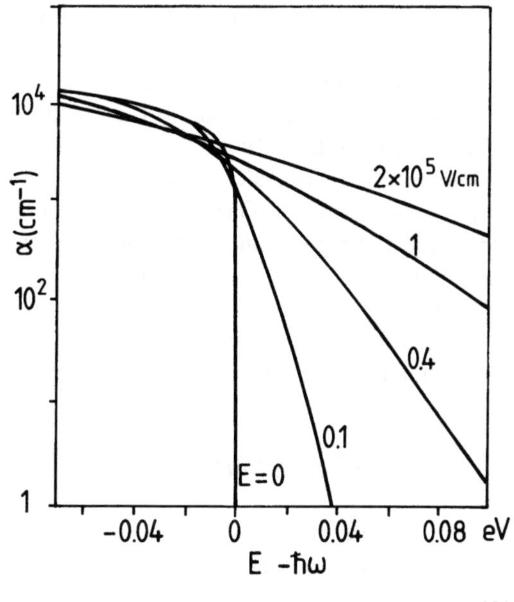

Fig. 5.17. Absorption coefficients α for GaAs at various electric fields E according to Ref. 187

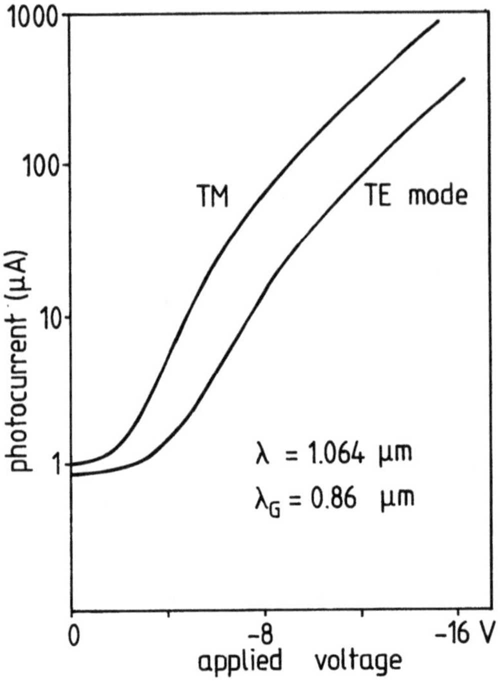

Fig. 5.18. Photocurrent in $Al_xGa_{1-x}As$-heterostructure[190] as a function of an electric field

advantage over silicon devices due to the wider bandgap and the higher electron mobility in GaAs[196]. Kingston et al.[194] have fabricated a GaAs CCD with every third electrode being transparent. Illuminating the electrodes with a narrowband radiation with a wavelength slightly longer than the direct-gap cut off wavelength the transmitted intensity depends on the signal charge under the transparent gate. When it is zero the field is high and the transmission is lowest because of electroabsorption. Thus the device works as a spatial light modulator which can be applied to coherent optical Fourier transformation[197].

In recent years efforts have been concentrated on the application of superlattices as modulators. Electroabsorption in these materials is not significantly different from the effect in bulk crystals. Due to the smaller dimensions modulation is not as effective as in bulk materials. In GaAs/GaAlAs quantum wells a transmission modulation of less than 50%[198] and in InGaAs/GaAs superlattices of only 6.4%[199] has been observed. On the other side the small dimensions should allow faster devices than possible with bulk electroabsorptive materials. With electrical pulses 131 ps modulation has been achieved[200]. Inducing the modulation field by laser pulses even subpicosecond modulation is possible[201]. The main advantage of superlattice modulators is the compatibility with technology used for sources and detectors based on the same material.

5.5.2 *Optical Bistability*

Bistable devices offer new properties which are not available with simple modulators. The basic idea is a feedback of the output intensity on the modulation voltage of an electrooptic crystal. In Fig. 5.19 the function of such a device is illustrated with a Mach-Zehnder interferometer. Assuming equal intensity in both arms the transmission characteristic is expressed by

$$I_{out} = I_{in}(1 + \cos\phi)/2 \qquad \text{with}$$

$$\phi = \frac{\pi}{\lambda} n^3 r \frac{L}{d} (V_m + V_b) ,$$

V_m being the modulation voltage and V_b the bias voltage. V_m depends on the outcoming light power:

$$V_m = \varkappa \cdot I_{out} .$$

Introducing normalized voltages by the use of Eq. (5.2) we obtain

$$I_{out} = I_{in}\left[1 + \cos\left(\pi \frac{\varkappa}{V_\pi} I_{out} + \pi \frac{V_b}{V_\pi}\right)\right]/2 . \tag{5.14}$$

This is a strongly nonlinear relation between I_{out} and I_{in}. Assuming the values $\frac{\varkappa}{V_\pi} = 1$ per unit of intensity and $\frac{V_b}{V_\pi} = 0.5$, Eq. (5.14) is illustrated in Fig. 5.20. For sufficient large input intensities several solutions of Eq. (5.14) exist. The intensity which is observed in an experiment shows a step-like behaviour[202] possibly accompanied by a hysteresis.

Equation (5.14) is a very simple expression describing bistability. In practice the features of a device also depend on its characteristic time responses. The dynamical behaviour of an electrooptic bistable device has been analyzed by Neyer and Voges[203]. They were able to realize experimentally stable, bistable, periodic, higher periodic, and chaotic solutions by using an integrated $LiNbO_3$-Mach-Zehnder interferometer with a delayed feedback.

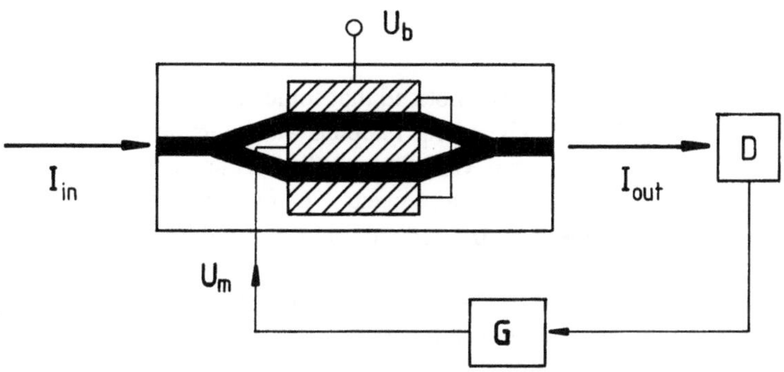

Fig. 5.19. Schematic arrangement of an electrooptic bistable Mach-Zehnder interferometer. *D* detector, *G* amplifier, U_b, U_m bias and modulation voltage

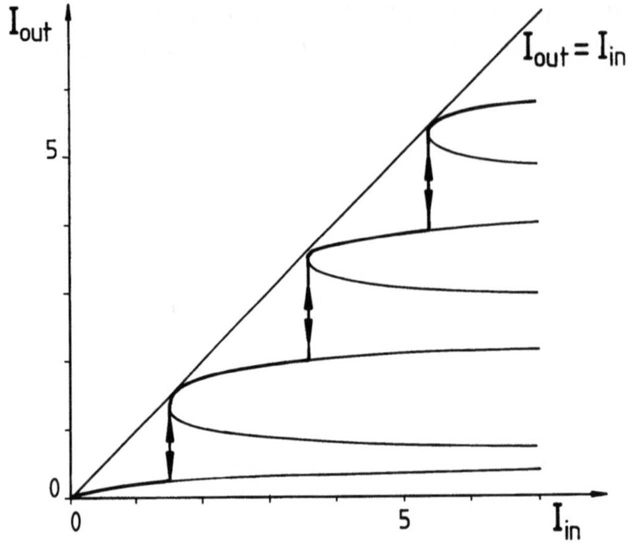

Fig. 5.20. Nonlinear relation between I_{in} and I_{out} in a bistable device

The first optical bistable device with feeding back an electrooptic crystal employed a Fabry-Perot resonator for switching a laser beam with switching powers of less than 1 μW[204]. Garmire et al.[205] used the electrooptically induced birefringence of a $LiNbO_3$-rod, placed between crossed polarizers, as a nonlinear medium. The first integrated bistable optical device without any source of electrical energy has been reported by Smith et al.[206]. They feeded a $LiNbO_3$-sample, placed in a Fabry-Perot cavity with the voltage produced by the photovoltaic detector. Aksenov et al.[207] utilized the total internal reflection at an electrooptical modulated interface. Sohler[208] has demonstrated with an integrated cutoff modulator, manufactured on the basis of $LiNbO_3$-waveguide, that it is possible to apply optical bistability as an electrooptical multivibrator. A theoretical analysis of self-oscillations has been reported by Laulicht[209] and their applications to optical communication systems by Schlaak and Kersten[210]. Optically bistable switches have also been demonstrated on the basis of multiple quantum well structures[211].

5.5.3 Photorefraction

In the late 1960's it had been noticed that some crystals are subjected to a special damage by laser radiation[212]. The effect happens at certain frequencies of light waves and is characterized by a spatial variation of the refractive index. Following investigations revealed that two different events occur. The first one is the generation of photo-induced currents and voltages and the second one the usual linear electrooptical effect.

If an absorbing crystal without a center of symmetry is illuminated with sufficient intensity, the ionization of electrons cause a photovoltaic current. Its density j is proportional to the absorbed power density:

$$j = \varkappa \alpha I \tag{5.15}$$

where α is the absorption coefficient and $\varkappa$ a constant. Fridkin and Magomadov[213] and Festl et al.[214] have verified the tensorial anisotropy of the photovoltaic effect. Since it is linear in the light intensity I, it must be bilinear in the unit polarization vector ϑ of the light wave and the tensor Eq. (5.15) can be written

$$j_i = \beta_{ijk}\vartheta_j\vartheta_k I \,.$$

Due to the absorption and nonuniform illumination the current density is not constant within the crystal. As a consequence electric fields are built up which generate a drift current in such a way that the sum of drift and photovoltaic currents is free of divergence[214]. Stopping the illumination, an inhomogeneous distribution of charge carriers, reflecting the original light intensities, is left. The resultant local electric fields modulate the index of refraction through the electrooptic effect. Now the spatially modulated crystal can be scanned by a second light beam. Its phase fronts vary accordingly to the electric fields induced by the first beam.

In detail the mechanisms of writein, readout, storage, and erase depend on the used material. Efficient photorefractive crystals, already applied as spatial light modulators, are ZnS[215,216], $Bi_{12}SiO_{20}$ (BSO)[216,217], DKDP[218], $BaTiO_3$[219], $SrBaNb_5O_{15}$ (SBN)[220], and $LiNbO_3$[221]. ZnS, BSO, $BaTiO_3$ and SBN are intrinsic photovoltaic crystals while $LiNbO_3$ is doped with Fe or Cu[222]. KDP and DKDP are addressed with an electron beam or the device is manufactured as a hybrid element, consisting of the electrooptic crystal and a suitable photoconductor. Such a hybrid device is shown in Fig. 5.21.

Casanent[224] has reviewed different kinds of spatial light modulators, which are not necessarily based on the electrooptical effect. A more recent and very comprehensive review has been reported by Günther[221]. One of the most important devices is the so called "Pockels readout optical modulator" (PROM), fabricated with a $Bi_{12}SiO_{20}$ (BSO)-crystal. The required voltage of about 2 KV for maximal modulation is comparatively large, but the advantage of BSO is its high sensitivity. At 441 nm only 0.1 J/cm^2 are needed for the write-in-process. At 633 nm the absorption of BSO is low enough for reading-out to be performed with a He-Ne-laser. To erase a stored pattern, the applied voltage is removed and the crystal is flooded with UV or blue light[210]. In the PROM

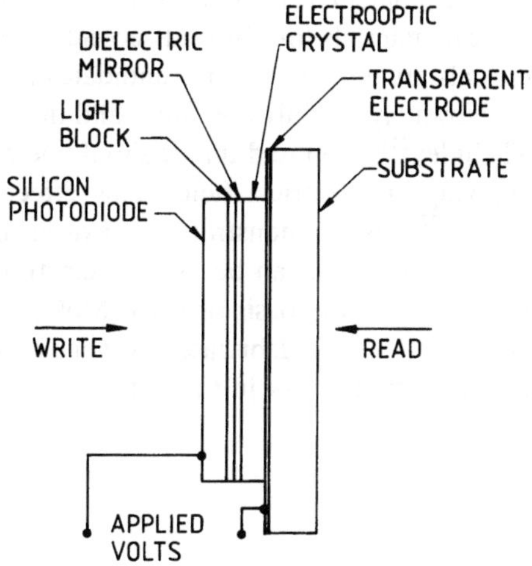

Fig. 5.21. Spatial light modulator after Ref. 223

modulator the longitudinal electrooptical effect is used, thus the orientation of the faces is (100). With (111)- and (110)-faces only transversal components of the field influence the light, such a device is called a PRIZ[225].

Casasent et al.[226] have tested and compared the U.S. PROM with the Soviet PROM and PRIZ. With the PRIZ-device a larger diffraction efficiency and higher spatial frequencies have been achieved. Since then Owechko and Tanguay[227] have made some theoretical efforts to push forward the understanding of spatial light modulators. Jonathan et al.[228] investigated the effect of externally applied electric fields on the buildup and decay of photorefractive gratings in BSO. Cronin-Golomb et al.[229] have considered the theory and applications of four-wave mixing in photorefractive crystals, showing new possibilities for their use.

6 Final Remarks

The speed of optoelectronic devices has steadily increased over the years. With the generation of electric pulses by ultrashort light pulses the detection of signals with sub-picosecond response times became necessary. Sampling oscilloscopes can resolve risetimes of only 25 ps. In the last years it has been shown that electrooptic sampling is a very powerful technique for making such time-domain measurements. The first experiment by Valdmanis et al.[141] is already described in Chap. 5.2.2. Recent developments have been reviewed by Kolner and Bloom[230]. This is not the only way to utilize the electrooptical effect in measurement techniques. Zhu et al.[231] have demonstrated electrooptic probing of electric fields in a GaAs coplanar waveguide sample which had a complex electrode structure. Obviously the stated examples represent only the beginning of a process; further applications of the electrooptical effect can be expected.

In the present review only technical applications are described. Electrooptical effects are also applied in basic research. Good examples are the study of the alignment of OH^--dipoles[232] and the spectroscopy of off-center impurities[233] in alkali-halides.

Often new developments in application of crystals are stimulated by new materials. At the moment $KTiOPO_4$ (KTP)[234] and β-BaB_2O_4[235] are discussed as new nonlinear optical materials for efficient second-harmonic generation. It can be expected that these crystals are also promising candidates for electrooptical applications.

Nowadays mainly the linear and the quadratic electrooptical effect are used. In 1965 Zheludev[236] proposed that it should be possible to create optical activity in a crystal by applying an electrical field. This effect is called electrogyration. Weber and Haussühl[11, 237] have demonstrated its existence in centrosymmetric crystals. Thus a different class of materials can be used in electrooptic modulation. Recently an electrogyration modulator on the basis of $NaBi(MoO_4)_2$ bulk crystals has been demonstrated[238]. Until now no attempt to fabricate a waveguide-electrogyration-device has been reported. This may be a future task in the field of electrooptics.

7 References

1. G. Szivessy: Kristalloptik: Handbuch der Physik XX, ed. H. Geiger and K. Scheel
   G. N. Ramachandran, S. Ramaseshan: Crystal Optics. Handbuch der Physik XX V/1 (ed. S. Flügge) Springer, Berlin 1961
   M. Born: Optik. Springer, Berlin 1972
   M. Born, E. Wolf: Principles of Optics. Pergamon Press, Oxford 1970
2. J. F. Nye: Physical Properties of Crystals. Clarendon Press, Oxford 1967
   T. S. Narasimhamurty: Photoelastic and Electro-Optic Properties of Crystals. Plenum Press, New York 1981
   S. Haussühl: Kristallphysik. Physik-Verlag, Weinheim 1983
3. S. Pancharatnam: Proc. Ind. Acad. Sci. *A46,* 1 (1957)
4. S. Pancharatnam: Proc. Ind. Acad. Sci. *A46,* 280 (1957)
5. R. C. Jones: J. Opt. Soc. Am. *31,* 488 (1941); *31,* 493 (1941); *31,* 500 (1941); *32,* 486 (1942); *37,* 107 (1947); *37,* 110 (1947); *38,* 671 (1948); *46,* 126 (1956)
6. A. Yariv, J. F. Lotspeich: J. Opt. Soc. Am. *72,* 273 (1982)
7. S. Haussühl, H.-J. Weber: Z. Kristallogr. *145,* 257 (1977)
8. A. Yariv, P. Yeh: Optical Waves in Crystals. John Wiley and Sons, New York 1984
9. L. Bohatý, J. Liebertz: Z. Kristallogr. *159,* 272 (1982)
10. J. Zyss: J. Non-Cryst. Solids *47,* 211 (1982)
11. H.-J. Weber, S. Haussühl: Phys. Stat. Sol. (b) *65,* 633 (1974)
12. J. C. Phillips: Phys. Rev. Lett. *20,* 550 (1968)
    J. A. Van Vechten: Phys. Rev. *182,* 891 (1969); *187,* 1007 (1969)
13. B. F. Levine: Phys. Rev. *B7,* 2600 (1973)
14. Ch.-Ch. Shih, A. Yariv: Phys. Rev. Lett. *44,* 281 (1980)
15. L. Bohatý: Z. Kristallogr. *166,* 97 (1984)
16. R. C. Miller, S. C. Abrahams, R. L. Barns, J. L. Bernstein, W. A. Nordland: Solid State Commun. *9,* 1463 (1971)
17. J. Minhua, H.-J. Weber: Z. Kristallogr. *154,* 127 (1981)
18. T. R. Sliker: J. Opt. Soc. Am. *54,* 1384 (1964)
19. H.-J. Weber: Acta Cryst. *A35,* 225 (1979)
20. P. Pellat-Finet: Optic Commun. *50,* 275 (1984)
21. I. P. Kaminow: Appl. Phys. Lett. *2,* 41 (1963)
22. I. P. Kaminow, T. J. Bridges, M. A. Pollack: ibid. *16,* 416 (1970)
23. I. P. Kaminow: ibid. *7,* 123 (1965); Erratum: ibid. *8,* 54 (1966)
24. O. E. De Lange: IEEE Spectrum *5,* October 77 (1968)
25. A. R. Johnston: Appl. Phys. Lett. *7,* 195 (1965)
26. H. Veeser, U. Bogner, W. Eisenmenger: Phys. stat. sol. (a) *37,* 161 (1976)
27. D. F. Nelson, E. H. Turner: J. Appl. Phys. *39,* 3337 (1968)
28. Landolt-Börnstein: Numerical Data and Functional Relationships in Science and Technology. Group III, Vol. 11 and 18. Springer, Berlin 1979 and 1984
29. S. K. Kurtz, T. T. Perry: J. Appl. Phys. *39,* 3798 (1968)
30. B. L. Davydov, L. D. Derkacheva, V. V. Duniva, M. E. Zhabotinskii, V. F. Zolin, L. G. Koreneva, M. A. Samokhina: Opt. Spectrosc. *30,* 274 (1971)
31. G. F. Lipscomb, A. F. Garito, R. S. Narang: J. Chem. Phys. *75,* 1509 (1981)
32. G. F. Lipscomb, A. F. Garito, R. S. Narang: Appl. Phys. Lett. *38,* 663 (1981)
33. M. E. Lines, A. M. Glass: Principles and Applications of Ferroelectrics and Related Materials. Clarendon Press, Oxford 1977
34. A. Kapenieks, M. Ozolinsk: Optik *63,* 333 (1983)
35. Landolt-Börnstein: Numerical Data and Functional Relationships in Science and Technology. New Series, Group III, Vol. 16a, Springer, Berlin 1981
36. S. A. Fedulov, Z. I. Shapiro, P. B. Ladyzhinskii: Sov. Phys. Crystallogr. *10,* 218 (1965)
    A. A. Ballman: J. Am. Ceram. Soc. *48,* 112 (1965)
37. K. Nassau, H. J. Levinstein, G. M. Loiacono: J. Phys. Chem. Solids *27,* 983 (1966)
    K. Nassau, H. J. Levinstein, G. M. Loiacono: ibid. *27,* 989 (1966)
38. B. C. Grabmaier, F. Otto: J. Crystal Growth *79,* 682 (1986)

39. P. K. Tien, A. A. Ballman: J. Vac. Sci. Technol. *12,* 892 (1975)
40. R. V. Schmidt, I. P. Kaminov: Appl. Phys. Lett. *25,* 458 (1974)
41. J. L. Jackel, V. Ramaswamy, S. P. Lyman: ibid. *38,* 509 (1981)
42. S. Yamada, M. Minakata: Jap. J. Appl. Phys. *20,* 733 (1981)
43. Standards on Piezoelectric Crystals, 1949. Proc. IRE 1378 (1949)
44. K. F. Hulme, P. H. Davies, V. M. Cound: J. Phys. C (Solid State Phys.) *2,* 855 (1969)
45. J. D. Zook, D. Chen, G. N. Otto: Appl. Phys. Lett. *11,* 159 (1967)
46. K. Onuki, N. Uchida, T. Saku: J. Opt. Soc. Am. *62,* 1030 (1972)
47. E. Bernal, G. D. Chen, T. C. Lee: Phys. Lett. *21,* 259 (1966)
48. P. V. Lenzo, E. G. Spenger, K. Nassau: J. Opt. Soc. Am. *56,* 633 (1966)
49. H. Iwasaki, H. Toyoda, N. Niizeka, H. Kubota: Jap. J. Appl. Phys. *6,* 1419 (1967)
50. E. H. Turner: Appl. Phys. Lett. *8,* 303 (1966)
51. I. P. Kaminov, E. H. Turner, R. L. Barns, J. L. Bernstein: J. Appl. Phys. *5,* 4379 (1980)
52. L. P. Avakyants, D. F. Kiselev, N. N. Shchitov: Sov. Phys. Solid State *18,* 899 (1976)
53. P. V. Lenzo, E. H. Turner, E. G. Spencer, A. A. Ballman: Appl. Phys. Letters *8,* 81 (1966)
54. H. Bach, J. Liebertz: Fortschr. Miner. *55,* 59 (1977)
55. P. B. Jamieson, S. C. Abrahams, J. L. Bernstein: J. Chem. Phys. *50,* 4352 (1969)
56. E. A. Giess, B. A. Scott, B. L. Olson, G. Burns, D. F. O'Kane: J. Am. Ceram. Soc. *53,* 14 (1970)
57. B. A. Scott, E. A. Giess, D. F. O'Kane: Mat. Res. Bull. *4,* 107 (1969)
58. J. J. Rubin, L. G. Van Mitert, H. J. Levinstein: J. Cryst. Growth *1,* 315 (1967)
59. G. Burns, D. F. O'Kane: Phys. Lett. *28A,* 776 (1969)
60. Yu. L. Kopylov, V. B. Kravchenko: Sov. Phys. Tech. Phys. *27,* 516 (1982)
61. S. Singh, D. A. Draegert, J. E. Geusic: Phys. Rev. *B2,* 2709 (1970)
62. T. Y. Tien, R. M. Garretson: J. Cryst. Growth *16,* 177 (1972)
63. S. H. Wemple, M. Di Domenico, I. Canlibel: J. Phys. Chem. Solids *29,* 1797 (1968)
64. V. Belruss, J. Kalnajs, A. Linz, R. C. Folweiler: Mat. Res. Bull. *6,* 899 (1971)
65. S. H. Wemple: Phys. Rev. *A137,* 1575 (1965)
66. J. J. Hurst, A. Linz: Mat. Res. Bull. *6,* 163 (1971)
67. H. Unoki, T. Sakudo: J. Phys. Soc. Jap. *21,* 1730 (1966)
68. T. Fukuda, Y. Mematsu: Jap. J. Appl. Phys. *11,* 163 (1972)
69. T. Fukuda, Y. Uematsu, T. Ito: J. Cryst. Growth *24/25,* 450 (1974)
70. H. Fay: Mat. Res. Bull. *2,* 585 (1967)
71. A. F. Devonshire: Phil. Mag. *40,* 1040 (1949); ibid. *42,* 1065 (1951)
72. J. E. Geusic, S. K. Kurtz, L. G. Van Mitert, S. H. Wemple: Appl. Phys. Lett. *4,* 141 (1964)
73. M. Di Domenico, Jr., S. H. Wemple: J. Appl. Phys. *40,* 720 (1969)
74. S. H. Wemple, M. Di Domenico, Jr.: ibid. *40,* 735 (1969)
75. J. C. Baumert, C. Walther, P. Buchmann, H. Kaufmann, H. Melchior, P. Günter: Appl. Phys. Lett. *46,* 1018 (1985)
76. J. Min-hua, F. Chang-shui, Y. Xi-lin, W. Min, Z. Tao-hong, G. Zhang-shou: J. Cryst. Growth *53,* 283 (1981)
77. Yu. V. Pisarevskii, G. A. Tregubov, Yu. V. Shaldin: Sov. Phys. Solid State *7,* 530 (1965)
78. H. Pursey, P. A. Merran, M. J. P. Musgrave: Brit. J. Appl. Phys. *18,* 285 (1967)
79. R. D. Rosner, E. H. Turner, I. P. Kaminow: Appl. Optics *6,* 778 (1967)
80. A. S. Vasilevskaya, A. S. Sonin: Sov. Phys. Crystallogr. *14,* 622 (1970)
81. E. N. Volkova, I. A. Velichko: ibid. *18,* 256 (1973)
82. J. H. Ott, T. R. Sliker: J. Opt. Soc. Am. *54,* 1442 (1964)
A. S. Vasilevskaya: Sov. Phys. Crystallogr. *11,* 644 (1967)
K. V. Rao, T. S. Narasimhamurty: J. Phys. C: Solid State Phys. *11,* 2343 (1978)
83. R. Onaka, H. Ito: Bull. Acad. Sci. USSR, Phys. Ser. *41,* 59 (1977)
84. J. L. Shay, J. H. Wernick: Ternary Chalcopyrite Semiconductors: Growth. Electronic Properties, and Applications. Pergamon Press, Oxford 1975
85. Landolt-Börnstein: Numerical Data and Functional Relationships in Science and Technology. New Series, Vol. 17a and 17b, Springer, Berlin 1982
86. A. J. R. De Kock: Crystal Growth of Bulk Crystals: Purification, Doping and Defects. Handbook of Semiconductors, Vol. 3. North-Holland Publishing Company, Amsterdam 1980
87. W. Wardzynski: Proc. Roy. Soc. *200A,* 370 (1961)
H. J. Skimin, D. G. Thomas: J. Appl. Phys. *33,* 56 (1962)

M. Inoue, I. Teramoto, S. Takayanagi: ibid. *33*, 2578 (1962)
N. R. Kyle: J. Electrochem. Soc. *118*, 1790 (1971)
88. A. G. Fischer: Methods of Growing Crystal under Pressure. Crystal Growth (ed. B. R. Pamplin) 521 (1975)
89. U. Zimmerli, A. Steinemann: J. Angew. Math. Phys. *16*, 555 (1965)
R. Gremmelmair: Z. Naturf. *11a*, 511 (1956)
90. Molecular Beam Epitaxy and Heterostructures (ed. L. L. Chang, K. Ploog) Martinus Nijhoff Publishers, Dordrecht 1985
91. D. A. B. Miller, D. S. Chemla, P. W. Smith, A. C. Gossard, W. T. Tsang: Appl. Phys. *B28*, 96 (1982)
92. H. Okamoto: J. Vac. Sci. Technol. *B3*, 687 (1985)
93. K. B. Kahen, J. P. Leburton: Appl. Phys. Lett. *47*, 508 (1985)
94. D. F. Nelson, E. H. Turner: J. Appl. Phys. *19*, 3337 (1968)
95. M. Sugie, K. Tada: Japan J. Appl. Phys. *15*, 421 (1976)
96. K. Tada, N. Suzuki: ibid. *19*, 2295 (1980)
97. J. E. Kiefer, A. Yariv: Appl. Phys. Lett. *15*, 26 (1969)
98. L. Baldassarre, A. Cingolani, M. Ferrara, M. Lugara: Solid State Commun. *34*, 237 (1980)
99. J. M. Ley: Electr. Lett. *2*, 394 (1966)
100. V. A. Shamburov, V. A. Kuznetsov, A. N. Lobacher, I. V. Kharitnova, V. G. Soshnikov: Sov. Phys.-Crystallogr. *15*, 252 (1970)
101. E. Mohler, B. Thomas: Phys. Stat. Sol. *b79*, 509 (1977)
102. L. Baldassarre, A. Cingolani, M. Sileilano: Opt. Commun. *26*, 72 (1978)
103. V. M. Cound, P. H. Davies, K. F. Hulme, D. Robertson: J. Phys. *C3*, L83 (1970)
104. H. Horinaka, H. Sonomura, T. Miyauchi: Japan J. Appl. Phys. *21*, 1485 (1982)
105. E. H. Turner, E. Buehler, H. Kasper: Phys. Rev. *B9*, 558 (1974)
106. M. Glick, F. K. Reinhart, G. Weimann, W. Schlapp: Appl. Phys. Lett. *48*, 989 (1969)
107. D. A. B. Miller, D. S. Chemla, T. C. Damen, A. C. Gossard, W. Wiegmann, T. H. Wood, C. A. Burrus: Phys. Rev. Lett. *53*, 2173 (1984)
108. T. H. Wood: Phys. Rev. Lett. *48*, 1413 (1986)
109. J. S. Weiner, D. A. B. Miller, D. S. Chemla, T. C. Damen, C. A. Burrus, T. H. Wood, A. C. Gossard, W. Wiegmann: ibid. *47*, 1148 (1985)
K. Wakita, Y. Kawamura, Y. Yoshikuni, H. Asahi, S. Uehara: IEEE J. Quantum Electron. *QE-22*, 1831 (1986)
110. I. P. Kaminow, E. H. Turner: Appl. Optics *5*, 1612 (1966)
111. F.-S. Chen: Proc. IEEE *58*, 1440 (1970)
112. S. H. Wemple, M. Di Domenico, Jr.: Applied Solid State Science *3*, 263 (1972)
113. P.-L. Liu: J. Appl. Phys. *53*, 6681 (1982)
114. R. Becker: IEEE J. Quantum Electron. *QE-20*, 723 (1984)
115. L. L. Steinmetz, T. W. Pouliot, B. C. Johnson: Appl. Optics *12*, 1468 (1973)
116. T. A. Kuzovka, E. V. Nilov, Yu. A. Flegontov: Opt. Spectrosc. *47*, 538 (1979)
N. P. Zanadvorov, V. A. Malinov, A. V. Charukkchev: ibid. *54*, 212 (1983)
117. J. Goldhar, M. A. Henesian: IEEE J. Quantum Electron. *QE-22*, 1137 (1986)
118. W. Demtröder: Laser Spectroscopy, Springer, Berlin 1981
119. A. Moritani, Y. Okuda, J. Nakai: Appl. Optics *22*, 1329 (1983)
120. E. A. Ballik, D. W. Lin: IEEE J. Quantum Electron. *QE-19*, 1166 (1983)
121. A. J. Rogers: Proc. I. E. E. *123*, 957 (1976)
122. C. Shaohuo: IEEE J. Quantum Electron. *QE-19*, 736 (1983)
123. R. C. Alferness: IEEE Trans. Microw. Th. Techn. *MTT-30*, 1121 (1982)
124. D. Marcuse: IEEE J. Quantum Electron. *QE-18*, 393 (1982)
125. F. Auracher, R. Keil: Wave Electron. *4*, 129 (1980)
126. P. Danielsen: IEEE J. Quantum Electron. *QE-20*, 1093 (1984)
127. I. P. Kaminow, V. Ramaswamy, R. V. Schmidt, E. H. Turner: Appl. Phys. Lett. *24*, 622 (1974)
128. I. P. Kaminow, L. W. Stulz, E. H. Turner: ibid. *27*, 555 (1975)
129. A. Neyer, W. Sohler: ibid. *35*, 256 (1979)
130. F. Auracher, R. Keil: ibid. *36*, 626 (1980)
131. F. J. Leonberger: Optics Lett. *5*, 312 (1980)
132. R. C. Alferness, N. P. Economou, L. L. Buhl: Appl. Phys. Lett. *38*, 214 (1981)

133. R. A. Becker: IEEE J. Quantum Electron. *QE-21,* 1144 (1985)
134. K. K. Wong, R. M. De La Rue, S. Wright: Optics Lett. *7,* 546 (1982)
135. H. Yajima, E. Sudo, J. Yumoto, K. Kashiwa: Appl. Phys. Lett. *45,* 214 (1984)
136. W. W. Rigrod, I. P. Kaminow: Proc. IEEE *51,* 137 (1963)
137. K. Kubota, J. Noda, O. Mikami: IEEE J. Quantum Electron. *QE-16,* 754 (1980)
138. C. M. Gee, G. D. Thurmond, H. W. Yen: Appl. Optics *22,* 2034 (1983)
139. R. A. Becker: Appl. Phys. Lett. *45,* 1168 (1984)
140. M. Izutsu, T. Sueta: IEEE J. Quantum Electron. *QE-19,* 668 (1983)
141. J. A. Valdmanis, G. A. Mourou, C. W. Gabel: ibid. *QE-19,* 664 (1983)
142. G. A. Mourou, K. E. Meyer: Appl. Phys. Lett. *45,* 492 (1984)
143. A. Yariv: IEEE J. Quantum Electron. *QE-9,* 919 (1973)
144. D. Marcuse: ibid. *QE-11,* 759 (1975)
145. S. Thaniyavarn: Appl. Phys. Lett. *47,* 674 (1985)
146. R. C. Alferness: ibid. *36,* 513 (1980)
147. R. C. Alferness, L. L. Buhl: Optics Lett. *5,* 473 (1980)
148. R. C. Alferness, L. L. Buhl: ibid. *7,* 500 (1982)
149. R. C. Alferness, L. L. Buhl: Appl. Phys. Lett. *47,* 1137 (1985)
150. D. Hall, A. Yariv, E. Garmire: Optics Commun. *9,* 403 (1970)
151. E. A. J. Marcatili: Bell Syst. Tech. J. *48,* 2071 (1969)
152. J. C. Campbell, F. A. Blum, D. W. Shaw, K. L. Lowley: Appl. Phys. Lett. *27,* 202 (1975)
153. M. Papuchon, Y. Combemale, X. Mathieu, D. B. Ostrowsky, L. Reibe, A. M. Roy, B. Sejourne, M. Werner: ibid. *27,* 289 (1975)
154. R. C. Alferness, N. P. Economou, L. L. Buhl: ibid. *38,* 214 (1981)
155. R. C. Alferness, C. H. Joyner, L. L. Buhl, S. K. Korotky: IEEE J. Quantum Electron. *QE-19,* 1339 (1983)
156. O. Mikami, J. Noda: Appl. Phys. Lett. *33,* 856 (1978)
157. R. V. Schmidt, H. Kogelnik: ibid. *28,* 503 (1976)
158. R. V. Schmidt, P. S. Cross: Optics Lett. *2,* 45 (1978)
P. S. Cross, R. V. Schmidt: IEEE J. Quantum Electron. *QE-15,* 1415 (1979)
159. S. I. Bozhevol'nyi, K. S. Buretskii, E. M. Zolotov, A. M. Prokhorov, V. A. Chernykh: Sov. J. Quantum Electron. *12,* 1165 (1982)
160. R. C. Alferness: Appl. Phys. Lett. *35,* 260 (1979)
161. R. C. Alferness: ibid. *35,* 748 (1979)
162. A. Kumar, A. N. Kaul, A. K. Ghatak: Optics Lett. *10,* 86 (1985)
163. L. A. Molter-Orr, H. A. Haus: Appl. Optics *24,* 1260 (1985)
164. J. P. Donnelly, N. L. Demeo, G. A. Ferrante, K. B. Nichols: IEEE J. Quantum Electron. *QE-21,* 18 (1985)
165. C. J. Setterlind, L. Thylen: ibid. *QE-22,* 595 (1986)
166. V. A. Zolotorev, A. V. Makovkin, V. L. Smitnov, V. P. Tarasov: Sov. J. Quantum Electron. *12,* 1241 (1982)
167. J. Finak, G. L. Yip: Optical and Quantum Electron. *17,* 15 (1985)
168. R. A. Forber, E. Marom: IEEE J. Quantum Electron. *QE-22,* 911 (1986)
169. G. E. Betts, W. S. C. Chang: ibid. *QE-22,* 1027 (1986)
170. L. Beiser: J. Opt. Soc. Am. *57,* 923 (1967)
171. W. Haas, R. Johannes, P. Cholet: Appl. Optics *3,* 988 (1964)
172. F. S. Chen, J. E. Gensic, S. K. Kurtz, J. G. Skinner, S. H. Wemple: J. Appl. Phys. *37,* 388 (1966)
173. R. P. Kiyatkin: Opt. Spectrosc. *38,* 209 (1975)
174. B. B. Sevruk, N. A. Gusak: ibid. *45,* 910 (1978)
175. B. B. Sevruk: ibid. *52,* 432 (1982)
176. I. P. Kaminow, L. W. Stulz: IEEE J. Quantum Electron. *QE-11,* 633 (1975)
177. Ch. S. Tsai, P. Saunier: Appl. Phys. Lett. *27,* 248 (1975)
178. Ch. S. Tsai, P. Saunier: Ferroelectr. *10,* 257 (1976)
179. J. F. Revelli: Appl. Optics *19,* 389 (1980)
E. T. Aksenov, A. A. Lipovskii, A. V. Pavlenko: Izv. Akad. Nauk SSSR, Ser. Fiz. *46,* 177 (1982)
180. Ch. L. Lee, J. F. Lee, J. Y. Huang: Appl. Optics *19,* 2902 (1980)
181. M. B. Chang: ibid. *21,* 3879 (1982)

182. Ch. S. Tsai, B. Kim, F. R. El-Akkari: IEEE J. Quantum Electron. *QE-14,* 513 (1978)
183. J. M. Hammer, W. Phillips: Appl. Phys. Lett. *24,* 545 (1974)
184. A. M. Belin, K. K. Swidzinskii, G. M. Turbina: Sov. J. Quantum Electron. *12,* 1153 (1982)
185. G. L. Tangonan, D. L. Persechini, J. F. Lotspeich, M. K. Barnoski: Appl. Optics *17,* 3259 (1978)
B. Chen, C. M. Meijer: Appl. Phys. Lett. *33,* 33 (1978)
186. A. M. Belin, V. N. Ryabokon, K. K. Svidzinskii: Sov. J. Quantum Electron. *10,* 249 (1980)
R. A. Becker, W. S. C. Chang: Appl. Optics *18,* 3296 (1979)
187. G. E. Stillman, C. M. Wolfe, C. O. Bozler, J. A. Rossi: Appl. Phys. Lett. *28,* 544 (1976)
188. T. E. Van Eck, L. M. Walpita, W. S. C. Chang, H. H. Wieder: ibid. *48,* 451 (1986)
189. R. H. Kingston: ibid. *34,* 744 (1979)
190. F. K. Reinhart: ibid. *22,* 372 (1973)
191. H. G. Bach, J. Krauser, H. P. Nolting, R. A. Logan, F. K. Reinhart: ibid. *42,* 692 (1983)
192. A. V. Maksimov, A. N. Blagodarov, N. A. Nagovitsyn, V. Ya. Kunir, V. I. Fomichev: Sov. Phys. Tech. Phys. *28,* 243 (1983)
193. G. E. Stillman, C. M. Wolfe, J. A. Rossi, J. P. Donnelly: Appl. Phys. Lett. *25,* 671 (1974)
194. R. H. Kingston, B. E. Burke, K. B. Nichols, F. J. Leonberger: ibid. *41,* 413 (1982)
195. W. S. Boyle, G. E. Smith: Bell System Tech. J. *49,* 587 (1970)
196. I. Deyhimy, R. C. Eden, J. S. Harris, Jr.: IEEE Trans. Electron Devices *ED-27,* 1172 (1980)
197. R. H. Kingston, F. J. Leonberger: IEEE J. Quantum Electron. *QE-19,* 1443 (1983)
198. T. H. Wood, C. A. Burrus, D. A. B. Miller, D. S. Chemla, T. C. Damen, A. C. Gossard, W. Wiegmann: Appl. Phys. Lett. *44,* 16 (1984)
199. T. E. Van Eck, P. Chu, W. S. C. Chang, H. H. Wieder: ibid. *49,* 135 (1986)
200. T. H. Wood, C. A. Burrns, D. A. B. Miller, D. S. Chemla, T. C. Damen, A. C. Gossard, W. Wiegmann: IEEE J. Quantum Electron. *QE-21,* 117 (1985)
201. W. H. Knox, D. A. B. Miller, T. C. Damen, D. S. Chemla, C. V. Shank, A. C. Gossard: Appl. Phys. Lett. *48,* 864 (1986)
202. D. A. B. Miller, S. D. Smith: Optics Commun. *31,* 101 (1979)
203. A. Neyer, E. Voges: IEEE J. Quantum Electron. *QE-18,* 2009 (1982)
204. P. W. Smith, E. H. Turner: Appl. Phys. Lett. *30,* 280 (1977)
205. E. Garmire, J. H. Marburger, S. D. Allen: ibid. *32,* 320 (1978)
206. P. W. Smith, I. P. Kaminow, P. J. Maloney, L. W. Stulz: ibid. *34,* 62 (1979)
207. E. T. Aksenov, A. V. Kukharev, A. A. Lipovskii, A. V. Pavlenko: Sov. Phys. Tech. Phys. *28,* 185 (1983)
208. W. Sohler: Appl. Phys. Lett. *36,* 351 (1980)
209. I. Laulicht: Optical and Quantum Electron. *13,* 259 (1981)
210. H. F. Schlaak, R. Th. Kersten: Optics Commun. *36,* 186 (1981)
211. D. A. B. Miller, J. E. Henry, A. C. Gossard, J. H. English: Appl. Phys. Lett. *49,* 821 (1986)
212. A. Ashkin, G. D. Boyd, J. M. Dziedtic, R. G. Smith, A. A. Ballman, J. J. Levinstein, K. Nassau: ibid. *9,* 72 (1966)
F. S. Chen: J. Appl. Phys. *38,* 3418 (1967)
213. V. M. Fridkin, R. M. Magomadov: JETP Lett. *30,* 686 (1979)
214. H. G. Festl, P. Hertel, E. Krätzig, R. von Baltz: Phys. stat. sol. (b) *113,* 157 (1982)
215. D. S. Oliver, P. Vohl, R. E. Aldrich, M. E. Behrendt, W. R. Buchan, R. C. Ellis, J. E. Genthe, J. R. Foft, S. L. Hou, G. McDaniel: Appl. Phys. Lett. *17,* 416 (1970)
216. J. Feinleib, D. S. Oliver: Appl. Optics *11,* 2752 (1972)
217. R. E. Aldrich, S. L. Hou, M. L. Harvill: J. Appl. Phys. *42,* 493 (1971)
S. L. Hou, D. S. Oliver: Appl. Phys. Lett. *18,* 325 (1971)
P. Nisenson, S. Iwasa: Appl. Optics *11,* 2760 (1972)
218. D. S. Sand, F. Faiman, W. J. Poppelbaum: IEEE J. Quantum Electron. *QE-9,* 708 (1973)
219. J. Feinberg, D. Heiman, M. R. Tanguay, Jr., R. W. Hellworth: J. Appl. Phys. *51,* 1297 (1980)
220. B. Fischer, M. Cronin-Golomb, J. O. White, A. Yariv, R. Neurgaonkar: Appl. Phys. Lett. *40,* 863 (1982)
221. P. Günter: Phys. Rep. *93,* 199 (1982)
222. R. A. Becker, R. C. Williamson: Appl. Phys. Lett. *47,* 1024 (1985)
223. D. Armitage, W. W. Anderson, T. J. Karr: IEEE J. Quantum Electron. *QE-21,* 1241 (1985)
224. D. Casasent: Proc. IEEE *65,* 143 (1977)

225. M. P. Petrov, V. I. Marakhonov, M. G. Shlyagin, A. V. Khomenko, M. V. Krasin'kova: Sov. Phys. Tech. Phys. *25,* 752 (1980)
226. D. Casasent, F. Caimi, A. Khomenko: Appl. Optics *20,* 4215 (1981)
227. Y. Owechko, A. R. Tanguay, Jr.: Optics Lett. *12,* 587 (1982)
Optics Commun. *44,* 239 (1983)
228. J.-M. C. Jonathan, R. W. Hellwarth, G. Roosen: IEEE J. Quantum Electron. *QE-22,* 1936 (1986)
229. M. Cronin-Golomb, B. Fischer, J. O. White, A. Yariv: ibid. *QE-20,* 12 (1984)
230. B. H. Kolner, D. M. Bloom: ibid. *QE-22,* 79 (1986)
231. Z. H. Zhu, J.-P. Weber, S. Y. Wang, S. Wang: Appl. Phys. Lett. *49,* 432 (1986)
232. S. Kapphan, F. Lüty: J. Phys. Chem. Solids *34,* 969 (1973)
233. S. E. Kapphan: Phys. stat. sol. (b) *113,* 147 (1982)
234. T. A. Driscoll, H. J. Hoffmann, R. E. Stone, P. E. Perkins: J. Opt. Soc. Am. *B3,* 683 (1986)
235. K. Kato: IEEE J. Quantum Electron. *QE-22,* 1013 (1986)
236. I. S. Zheludev: Sov. Phys.-Crystallogr. *9,* 418 (1965)
237. H.-J. Weber, S. Haussühl: Z. Kristallogr. *146,* 303 (1977)
238. E. Kostova, M. Kostova: Appl. Optics *24,* 1726 (1985)

Erratum

Crystals, Growth, Properties, and Applications
Vol. 10, "Growth and Defect Structures"

Re: "Hydrodynamics, Heat and Mass Transfer During Crystal Growth"
V. I. Polezhaev, pp. 87ff.

Corrections

Page 93 Diffusion Marangoni number: $\frac{\sigma_0 \beta_{0c} L}{\varrho v^2} \Delta C$

Diffusion Bio number: $\frac{K - 1}{D} V_f R$

Page 97 Eq. 8 and 9: $C = 1$ instead of $c = 1$

Page 98 Table 3:
H_c/R_c instead of v_c/R_c

Page 100 Paragraph 2.3.2:
$5\sqrt{vx/V}$ instead of $5\sqrt{vx/v}$
$Re_x V_x/v \sim 5 \cdot 10^5$ instead of $Re_x v_x/v \sim 5 \cdot 10^5$

Page 105 Fig. 5b inset read T_B and T_H instead of Θ_B and Θ_H
Figure caption: $\Delta\Theta_m = (T_B - T_H)/(T_2 - T_1)$

Page 116 Paragraph 3.1.1:
$K = 0$ instead of $\varkappa = 0$

Page 119 centre of page:
read: "conduction" instead of "heat transfer"

Page 121 Fig. 15 a–d caption
$Re = VH/v$

Page 122 $Pr = 0.02$ instead of $Pr = 0.2$

Page 123 Line 8 from the bottom: read "in a melt" instead of "in an indium melt"

Page 124 Below centre of page:
"$S = L - v_f t$ (Fig. 16a), where t is the time."
"boundary conditions" instead of "boundary layers"
"fluxes to all regions" instead of "fluxes fo all regions"
$L/H = 4$ instead of $H/R = 4$
$\frac{v_f H}{\nu} = 0.2$ instead of $\frac{V_f H}{\gamma} = 0.2$
Figure 16
$u = v = 0,\ T = T_c + \gamma(S - x),\ dc/dy = 0$
V_f instead of v_f
$D\ dc/dx = v_f(1 - K_0)C$

Page 125 Fig. 16c

y = 1

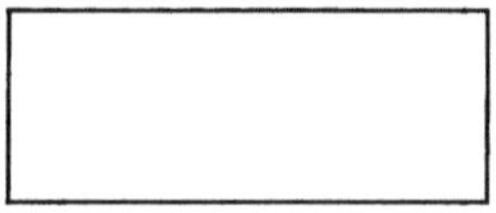

y = 0

Caption: Impurity distribution C′ along the crystal axis $D = 1.7 \cdot 10^{-4}$ cm²/S
■ – local concentration (y = 1)
$1 = \frac{L - S}{L} = 0.375$

Page 127 Figure 18a, b caption
$H/R_c = 1$ instead of $H/L = 1$

Page 130 Above centre of page: 0.1 cm instead of 0.1 sm

Page 131 $Gr_{RC} = (g\beta R_c^3/\nu^2)\Delta T$
Eq. 19 $D_s = (g\beta\Delta T R_S^3 \pi^{-2})^{1/4}\Omega_S^{-1/2}$

Page 138 Paragraph 4.2
$L/H \gg 1$ instead of L/H 1

Page 139 "For Ge and InSb melts" instead of "ZnSb"

Page 140 3rd line from the top
(Si, Se) instead of (Si, Sc)

Ref. 11 eds. Ueda, R., ...

Ref. 49 Lyumkis, E. D.

Ref. 86 Kirdyashkin, A. G.

Ref. 104 Kobayashi, R.

Ref. 118 Bannister, T. C.

Ref. 142 Fizika Atmosfery I Okeana

Ref. 207 Cockayne, B.

Ref. 219 J. Electrochem. Soc.

Author Index Volumes 1–11

The volume numbers are printed in italics

MIX
Papier aus verantwortungsvollen Quellen
Paper from responsible sources
FSC® C105338

If you have any concerns about our products,
you can contact us on
ProductSafety@springernature.com

In case Publisher is established outside the EU,
the EU authorized representative is:
Springer Nature Customer Service Center GmbH
Europaplatz 3, 69115 Heidelberg, Germany

Printed by Libri Plureos GmbH
in Hamburg, Germany